DYNAMIC DECISION THEORY:
APPLICATIONS TO URBAN AND REGIONAL TOPICS

STUDIES IN OPERATIONAL REGIONAL SCIENCE

Folmer, H., Regional Economic Policy. Measurement of its Effect. 1986. ISBN 90-247-3308-1.

Brouwer, F., Integrated Environmental Modelling: Design and Tools. 1987. ISBN 90-247-3519-X.

Toyomane, N., Multiregional Input–Output Models in Long-Run Simulation. 1988. ISBN 90-247-3679-X.

Anselin, L., Spatial Econometrics: Methods and Models. 1988. ISBN 90-247-3735-4.

Fotheringham, A.S. and O'Kelly, M.E., Spatial Interaction Models: Formulations and Applications. 1988. ISBN 0-7923-0021-1.

Haag, G., Dynamic Decision Theory: Applications to Urban and Regional Topics. 1989. ISBN 0-7923-0194-3.

Dynamic Decision Theory: Applications to Urban and Regional Topics

by

Günter Haag

Deutsche Physikalische Gesellschaft

KLUWER ACADEMIC PUBLISHERS
DORDRECHT / BOSTON / LONDON

Library of Congress Cataloging in Publication Data

Haag, G. (Günter), 1948-
Dynamic decision theory : applications to urban and regional topics / Günter Haag.
p. cm. -- (Studies in operational regional science ; 6)
Includes bibliographies and index.
ISBN 0-7923-0194-3 (U.S.)
1. Migration, Internal--Mathematical models. 2. Decision-making--Mathematical models. 3. Regional economics--Mathematical models.
I. Title. II. Series.
HB1952.H33 1989
304.6'0724--dc19 89-2407

ISBN 0-7923-0194-3

Published by Kluwer Academic Publishers,
P.O. Box 17, 3300 AA Dordrecht, The Netherlands.

Kluwer Academic Publishers incorporates
the publishing programmes of
D. Reidel, Martinus Nijhoff, Dr W. Junk and MTP Press.

Sold and distributed in the U.S.A. and Canada
by Kluwer Academic Publishers,
101 Philip Drive, Norwell, MA 02061, U.S.A.

In all other countries, sold and distributed
by Kluwer Academic Publishers Group,
P.O. Box 322, 3300 AH Dordrecht, The Netherlands.

printed on acid free paper

Printed in The Netherlands

Contents

Preface

Choice processes appear in all spheres of society. Hitherto ruling paradigms in the modelling of choice problems have presumed a competitive general equilibrium which, however, proves insufficient for dynamic processes.

This contribution aims at providing a general coherent and closed framework for the dynamic modelling of decision processes. It was one of my main interests to build a bridge between the pure model building concepts and their practical applications. Therefore all given examples are related to empirical work.

Solution algorithms for the estimation of trend parameters as well as the numerical simulation in concrete applications therefore play a central role in this contribution.

Friendly relations with a number of colleagues from many universities in Europe, and the U.S. have emerged during the different applications. I wish to thank all of them. The international cooperations were mainly initiated and supported by conferences and workshops organized and financed by the *International Institute for Applied Systems Analysis* (IIASA), the *Istituto Ricerche Economico-Sociali Del Piemonte* (IRES), the *Institut National D'Etudes Démographiques* (INED), the *Centre for Regional Science Research Umeå* (CERUM) and the *Projets de Cooperation et D'Echange avec France* (Procop). Special thanks go to the Volkswagen *Stiftung* for financial support of this work over the years.

Thanks also go in particular to my friend and mentor Prof.W.Weidlich for his encouragement and for the many suggestions he made in fruitful discussions and common work that have taken place over the years.

Finally I am very indebted to Prof. P. Nijkamp for his interest in this work and for making many valuable suggestions.

My sincere thanks also go to Mrs. W. Brown, who did the editorial work of the manuscript with great patience and care.

Stuttgart, December 1988 Günter Haag

Chapter 1

Introduction

Our theory is thoroughly static. A dynamic theory would unquestionably be more complete and, therefore, preferable. But there is ample evidence from other branches of science that it is futile to try to build one as long as the static side is not thoroughly understood . . .

Our static theory specifies equilibria . . . A dynamic theory, when one is found - will probably describe the change in terms of simpler concepts.

J. von Neumann

This quotation of Von Neumann can be found in his book: *Theory of Games and Economic Behaviour,*[1] written in 1944 in collaboration with Morgenstern. Indeed, research has focused in the last decades on the strong understanding of the static situation of socio-economic systems. However, it is now technically possible to begin to fulfil Von Neumann's dream (see Aubin et al[2]). We are more and more becoming aware of the important role that unstable and chaotic motion plays in the dynamics of macrosystems.

Choice problems are concerned with *agents* (individuals and firms to mention a few) who have to select one alternative from a set of alternatives. On the basis of static considerations such choice models have already been developed. In the Appendix, one of the most popular models, namely the multinomial logit model, is presented. However, all these approaches are limited in their usefulness since the time factor, i.e., the delay between causes and effects is excluded and most of the models also do not allow the introduction of social interactions among the individuals (de Palma[3]). The dynamic theory of decision processes presented here aims at providing such an adequate tool.

The evolution of a socio-economic system is not an autonomous process, but the result of human decisions, occuring over time as a broad stream of concurrent, unrelated or interrelated, individual or corporate choices. The mechanisms behind the millions of private decisions taken every day cannot be controlled and influenced by public authorities, at least not in a direct way.

Therefore, planners in charge of such systems face the difficult task of making decisions concerning a system which is largely subject to external influences in the form of national policies and entangled economies on the one hand, while on the other hand the system is controlled by decisions of private firms, investors, and other individual or corporate agents. Only limited instruments of policy are available and at their disposal, and it is of crucial importance to know in advance which of these are likely to be most effective.

In this contribution we examine search and choice behaviour of individual agents in an environment in which alternatives become feasible. Such processes may, for example, concern with households searching for dwellings, individuals seeking employment, and inter - and intra- regional migration of individuals. The approach presented in this book is based on a stochastic modelling framework, it holds under comparatively weak assumptions, and therefore enlarges the scope of model building in social and economic sciences. Each socio- economic system namely can be regarded as a multicomponent system consisting of groups of decision-makers orientating their activities with respect to the environment (the market).

This approach is related to *Synergetics* a new branch of science. Synergetics concentrates on the structural selforganizing space- time features of such multicomponent systems. Although the interactions and the constituting units of the various systems under consideration seem to be completely uncomparable on the microlevel there exists a close analogy between them on the macrolevel. The interdisciplinar universality of synergetics has its origin in the unifying concepts of model building and classification of such phenomena. The conceptual framework of Synergetics has been introduced by Haken.[4]

In natural sciences the elementary units and the fundamental interaction constituting the basic system are well known. In these sciences, in principle, model assumptions can directly be verified or falsified by experimental tests, and the reproducibility of experiments is fundamental and constitutive. Typically , one and the same experiment must and can be repeated under identical conditions in order to measure the value - or the statistical distribution of values - of an observable with a definite precision.

In social sciences, however, the interactions between the elementary units (such as individuals and firms for instance) are rather unknown and cannot be derived from first principles. Experimental tests are mostly impossible to repeat under identical socio-economic conditions. The empirical data base related to a certain subject is often rather limited and even the comparability among data sets of different countries concerning the same subject is not always guaranteed.

The macroscopically observable quantities denoted as *gross variable* or *macro variable* can be changed by varying certain *control parameters* of the system. Radical variations of the macroscopic state - denoted as *phase transitions* may occur if the control parameters pass certain *critical values* . *Order parameters* are suitable macro variables characterizing this change of order. Therefore, in a phase transition new order parameters may arise or decay. If a control parameter approaches the vicinity of a critical value for a phase transition, an enhancement of *fluctuations* can be observed until the fluctuations become macroscopic and trigger the system into a new phase. Therefore, these fluctuations are necessarily connected with the phase transition and in a sense anticipate the emergence of a new structure or the decay of an old one.

So far as natural sciences are concerned, much research has been done in investigating synergetic phenomena in the neighbourhood of physical science (see e.g., Haken,[5] Pacault and Vidal,[6] Güttinger and Eikemeier,[7] and Blumenfeld [8]). However, the application of synergetic concepts to socio- economic processes is still a rather new field (Weidlich[9, 10]). Nevertheless, interest in this field is rapidly growing and intense interdisciplinary work at an international level has already started.

A framework for modelling a wide class of socio-economic phenomena has been given in Weidlich and Haag.[11] In the following we will partially proceed along the line of argumentation given in this book.

Coming to the comparison between social sciences with natural sciences, some critical remarks must be made at the beginning: first, there exists no direct *short-cut* to transfer concepts from the natural sciences to social sciences. Appropriate and characteristic concepts have to be developed for the quantitative description of socio-economic processes. Second, synergetics can be applied only under certain conditions to a certain class of genuine social phenomena. If these conditions are fulfilled, however, a true structural relationship between natural and social sciences and not an accidental analogy will have been found.

Human society can be regarded as a multi-component system whose members, the individuals, adopt different attitudes or kinds of behaviour. The causes of global changes in society are assumed to be correlated to the decisions of individuals to change their attitudes. In contrast to physico-chemical systems, however, the elementary units (agents) and the interactions within human society are of a rather complex nature. A complex mixture of fluctuating rational considerations, professional activities and emotional preferences and motivations finally merge into one of relatively few well demarcated resultant attitudes. These attitudes may be in the field of religion, politics, education, habitation, occupation, economic standard, consumer habit, family status, sport activity etc. The manifold of possible attitudes, the *attitude space*, is an open one, since hitherto unknown attitudes may develop.

Due to individual decision processes caused by experience, eliciting thoughts and emotions, transitions from one attitude to another are possible. On the macrolevel, in turn, the *decision configuration* describes the distribution of attitudes of a socio-economic system and may be considered as an appropriate set of macro variables for the system under consideration.

However, the microlevel of details of the complex interplay of rational and emotional, conscious and subconscious, genetic and environmental influences on the decisions of individuals is unknown. Hence, a probabilistic description instead of a deterministic explication of the decision process seems to be adequate. Therefore, the probability that a certain decision configuration is realized will be introduced. The *master equation* is the equation of motion for this probability distribution. The essential constituent of this equation is the *transition rate* for a member of a considered subpopulation to change from one attitude to another. The modelling of such transition rates in terms of socio-economic variables will turn out to be the central part of model building in socio-economic systems.

The probability distribution over a given decision configuration contains the most detailed information about the system. In particular not only the mean values (or sometimes the most probable values) of certain attitudes can be calculated but also their mean square deviations. Correspondingly, the amount of mathematics to solve the time-dependent master equation is considerable. However, in most cases, the full information contained in the probability distribution cannot be exploited because of lack of sufficiently comprehensive empiric data.

Therefore, it is indicated to make a transition to a less exhaustive description in terms of quasi-closed equations of motion for the mean values and variances only. These equations of motion can be derived from the master equation in a straightforward manner. Hence, the master equation provides the link between the microscopic concept of individual decision processes and the macroscopic equations of motion for aggregate mean values.

The generally nonlinear form of the equations of motion expresses the structure of self-consistence prevalent in all socio-economic systems, namely a *cyclic coupling* between causes and effects in society:

Through their cultural and economic activities the individual members of a society contribute to cultural, political, religious, social and economic components. This collective field determines the socio-political atmosphere and the cultural and economic standard of society. Therefore, this collective field acts as an order parameter of the socio-economic system and characterizes the phase in which the society exists.

On the other hand, the collective field strongly influences the individuals in the society by orientating (manipulating) their activities. The feedback between individuals and the collective field - the cyclic coupling of causes and effects - may lead to a quasi-stable temporal development of the system characterized by a certain predictability of its trajectory. However, highly divergent alternative paths of evolution of society are possible if the control parameter of the system attains certain critical values. Fluctuations on a microscale - for instance, the actions of very few influential people - may decide into which of the diverging paths the society will bifurcate. The probability distribution for the decision configuration will have lost its simple unimodal structure.

Under these circumstances the forecasting of socio-economic variables becomes rather difficult if not impossible. On the macrolevel this problematic situation often results in terms of limit cycles or even chaotic behaviour of system variables. Examples which demonstrate such complicated dynamics will be given in Chapters 4 and 6. An excellent introduction into the field of chaotic systems is given by Haken.[12]

Up to now we have considered one of the main tasks in social sciences namely to provide the link between the microscopic level of decisions of individuals and the macroscopic sphere of the dynamics of aggregate mean values. However, this important link is not only of theoretical interest. It enables us to introduce a second important link, namely the one between the empirical observations and the theoretical model constructed on the basis of the first link. Of course, it is well known in this context that it is difficult, if not impossible, to give a direct and unique causal interpretation of the socio-economic situations and the behaviour of certain macrovariables of society in terms of individual motivations on the microlevel . Instead we expect that many of such motivations merge with different intensities in producing the observed macro-dynamics.

Since in social sciences the empirical data basis is often in a rather bad state, it is one model building criteria to introduce variables only which are empirically accessible - at least in principle.

The model parameters, denoted as *trend parameters* are then determined by minimizing the sum of the square deviations between theoretical and empirical expressions. In other words we apply the method of least squares.

The equations for the trend parameters obtained in this way can be solved either analytically or by applying different solution algorithms. In particular, Schwefel's evolutionary strategy [13] has demonstrated its advantage in parameter spaces of high dimension.

At the end of this chapter a few remarks are due concerning the role of mathematics in this theory: *nonlinear* equations of motion with only a few relevant parameters but a high variety of possible nontrivial solutions are preferred to a high-dimensional system of *linear* equations of motion with a great number of open parameters. Although this statement seems to be trivial at first, the construction of simple nonlinear models with clearly interpretable trend parameters is not that easy and needs professional experience in the social sector under consideration.

We shall now give a survey of the following chapters. In Chapter 2 the general concepts of the dynamic theory of decision processes will be introduced.

The method is applied to the problem of shocks in urban evolution in Chapter 3. After a short discussion of existing theories on this global phenomenon we present a simple model exhibiting various continuos or catastrophic shifts in population trajectories. Different outcomes can be attributed to different behavioural and technological circumstances.

In Chapter 4 a stochastic nonlinear dynamic model of intra-urban residential rent and density interactions is presented: individual buyers are considered who are willing to pay per unit area of land amounts depending among other factors on the current aggregate zonal residential density. On the other hand, individual suppliers of land are also introduced, interacting with the buyers on the land market, through an asking rent. The behaviour of both groups of agents is assumed to be stochastic. The deterministic mean value equations of aggregate behaviour are derived for the relative residential density and the rent in the zones. Finally the model is applied to twelve Standard Metropolitan Statistical Areas (SMSAs).

Some results of a stochastic theory of interregional migration are presented in Chapter 5. It is the main purpose of this chapter to demonstrate the evaluation of the mobility and the regional utilities and preferences using the migratory data. Furthermore, the selection and ranking of key-factors out of the available set of socio-economic variables is implemented.

We shall demonstrate in Chapter 6 that deterministic chaos can appear in our general migration model. The inter-group and intra-group interaction of individuals decisively influence the functional shape of the utility function of the individuals and determine the system dynamics. Contrary to Chapter 5 we consider here the migration of subpopulations and their socio-economic interaction.

In Chapter 7 we establish a dynamic service sector model with prices and land rents. Starting point is the decision processes of both the individuals who decide to use facilities of different regions and of the entrepreneurs who decide about changes in the facility stock. The model is concerned with four kinds of agents whose decisions control destination choice (consumer), facility size (developers), prices of goods or services (retailers) and land rent (land lords).

In Chapter 8 an attempt is made to provide a survey of further possible applications of our dynamic theory of decision processes. We want to open especially the large field of applications in economics and in political decision making.

The Appendix (Chapter 9) deals with the master equation and its embedding into the general theory of Markov processes. The most important properties of the master equation will be presented there as well as its stationary solution in the case of detailed balance. We will also discuss a number of properties of a multinomial logit model and the way in which this static approach is related to our theory.

Chapter 2

A Dynamic Theory of Decision Processes

While the individual man is an insoluble puzzle, in the aggregate he becomes a mathematical certainty. You can, for example, never fore- tell what any one man will do, but you can say with precision what an average number will be up to. Individuals vary, but percentages remain constant.

Sir A.C.Doyle

Over the past years, the use of disaggregate choice models has been strongly advocated (see Golledge and Timmermans,[1] Ben-Akiva and Lerman,[2] Bahrenberg, Fischer and Nijkamp,[3] Pitfield,[4] Johnson and Hensher [5]), for by using such models it is possible to capture stochastic and behavioural aspects of decision processes. New insights into decision-making and choice behaviour processes may be obtained by modelling at the level of individual actors in a system. For example a spatial system in which the actors are consumers or suppliers of activities such as migrants, travellers, property developers or local government decision-makers. Therefore, considerable efforts have been devoted to the development of behavioural spatial choice models capable of considering individual choices from a set of discrete alternatives at a point in time. With very few exceptions the emphasis on such discrete choice models has been strictly cross-sectional even if the choice processes studied were inherently dynamic in nature.

Quite recently increasing attention has been paid to modelling dynamic processes. The reasons for this focus are well known. They relate essentially to a concern of economic, social and environmental changes in general and to an interest in identifying the influences on change and understanding the dynamics of choice behaviour in particular. Over the last few years several approaches to modelling dynamic choice processes have been developed. An account of different views in dynamic choice processes is given by Fischer et al.[6] In this introductory chapter we will follow the argumentation given in this contribution. The different approaches differ widely in scope and in methodology. A major distinction among these approaches can be made with respect to the temporal unit of the analysis (continuous versus discrete). Correspondingly discrete-time and continuous-time dynamic approaches may be distinguished. Continuous-time approaches avoid the potentially arbitrary nature of the definition of the time unit in discrete-time approaches and make it possible to incorporate explicitly time in specific change points. Discrete-time approaches have to identify "natural" decision periods which are invariant across the population of sampled individuals. The parameters derived in the latter case are generally not invariant to the position of and the length between the time separation points. Discrete- and continuous-time approaches may be further disaggregated according to the nature of choice (discrete versus continuous choice). Thus, four broad types of approaches in modelling dynamic choice processes may be distinguished (see Fig. 2.1). Only very recently there have been attempts to integrate continuous and discrete choices intertemporally (see, for example, Hensher[7]).

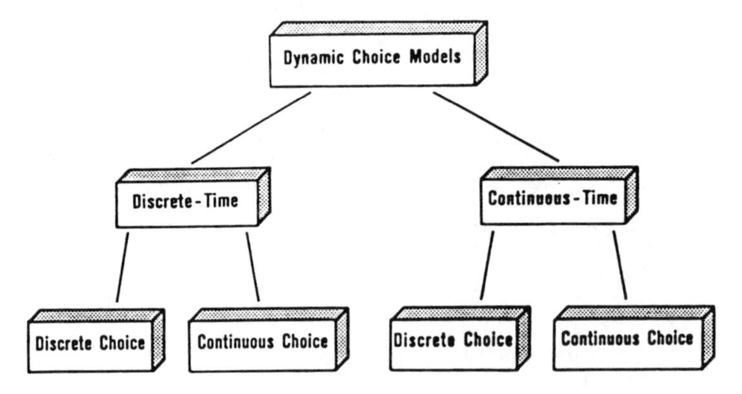

Figure 2.1: Different classes of dynamic choice modelling approaches

The emphasis in this book is on discrete-time and continuous-time discrete choice model approaches. First, the panel data-based discrete time discrete choice model approach will be briefly described. Then, the master equation approach will be introduced in detail for the modelling of continuous-time discrete choice processes. Moreover, its relationship to the (static) multinomial logit model will be shown.

2.1 The Panel Data-Based Discrete Choice Approach

In the recent past social and economic scientists have shown an increasing interest in the potential which longitudinal survey data offer to measure and model the components of behavioural change at the individual level (Coleman,[8] Tuma and Hannen,[9] Hensher and Wrigley,[10] Wrigley[11]). The essence of panel data is information on a (more or less) fixed sample of decision-makers during time so that statements can be made about the behavioural response at the individual level. Panel data may be obtained by classical panel surveys which involve repeated measurements on the same individuals at different points in time, by rotating panel surveys which are characterized by a process of planned "retirement" of sample units and systematic "refreshment" by new representative sample units, or by mixed panel surveys which are hybrids of classical panel surveys on the one hand and rotating panel surveys or repeated cross-sectional surveys on the other hand.

The great potential of panel data for dynamic modelling stems from the temporal nature of the data and the data linkage for each decision-maker. Panel data enable one to recognize explicitly the intertemporal nature of choice outcomes, especially the effect of experience on decisions. Moreover, it is expected that the use of panel data will result in greater efficiency, in both statistical and behavioural terms, than the estimation of separate relationships in the case of a repeated cross-sectional sample (Johnson and Hensher,[5] Coleman[8]). A major shortcoming of repeated cross-sectional surveys refers to the fact that the sample units are not retained from one time period to the next. There is no possibility to decompose observed change in behaviour over time into the two components: changes in population composition and changes in sample behaviour. Thus, dynamic models of discrete choice have to be based on panel data.

The critical issues in an intertemporal specification of a choice model are related to the proper treatment of three types of systematic variation: *heterogeneity, non-stationarity* and *structural state dependence*. Heterogeneity refers to the variation among individuals due to both observed and unobserved external influences where both may result from the censoring of the panel data base and/or from the omission of important time-varying or time-invariant influences during the sample period. This form of dependency may be treated in a number of different, but not necessarily mutually exclusive ways. For example, the set of decision-makers may be disaggregated by exogeneous characteristics or by decision process characteristics in order to account for heterogeneity or taste variations. Non-stationarity refers to the variation in individual and aggregate choice probabilities, resulting for example, from changes in the behavioural environment. The third type of variation, namely structural state dependency refers to the dependency of current individual choice probabilities on the preceding individual history. Structural state dependence effects may arise due to several reasons. Choice outcomes may depend on previous choices, on the period of time the current state has been occupied, on previous interchoice times (lagged duration-dependence effect) and on the number of times different states have been occupied (Wrigley[11]). For practical reasons it might be useful to assume that one or more of these sources of state dependence are unimportant and, thus may be neglected for the choice processes under consideration.

The methodological problem posed to the researcher by the presence of all three types of systematic variation in the data is very considerable. It is already not a task to disentangle the influences of intertemporal state dependence and heterogeneity, especially when some choice-relevant influences are unobserved (i.e., neglected or unmeasurable) and if they are temporally invariant and, thus, correlated with any time invariant observable variable. Moreover, omitted variables may and most likely do introduce a spurious time-dependence effect and bias into the parameter estimates of the observed exogeneous variables. It is clear that the identification of the three types of systematic variation and in particular of state dependence effects is of vital importance for satisfactorily modelling the dynamics in choice processes in the framework of a panel data-based context.

A general intertemporal representation of individual choice behaviour ideally requires including terms to represent all the dimensions of intertemporal behaviour (Hensher and Wrigley[10]), i.e.,

$$\begin{pmatrix}\text{current}\\ \text{choice}\end{pmatrix} = f\left\{\begin{pmatrix}\text{current, past and/}\\ \text{or future levels of}\\ \text{exogen. variables}\end{pmatrix}, \begin{pmatrix}\text{effects of the relevant}\\ \text{entire (or part) past}\\ \text{history}\end{pmatrix}, \begin{pmatrix}\text{cumulative of the most}\\ \text{recent continuous}\\ \text{experience in a state}\end{pmatrix}, \begin{pmatrix}\text{accounting for}\\ \text{heterogeneity}\end{pmatrix}, \begin{pmatrix}\text{habit}\\ \text{persistence}\end{pmatrix}\right\}$$

Heckman[12] developed a model which fulfils most of these requirements. His discrete time individual choice model is sufficiently flexible to take into account time-dependent explanatory variables and to account for complex structural state dependence inter-relationship and heterogeneity.

The model is based upon the following ideas. It is assumed that from a random sample of choice makers or individuals information on the presence or absence of an event (i.e., choice outcome) in each of T equi-spaced time periods is assembled. The key assumption of the model is that discrete outcomes are generated by continuous variables with cross-thresholds, or more precisely that an event for decision-maker i in time period t occurs, if and only if a continuous latent random variable y_{it} crosses a threshold. In applications, such variables are related to well defined economic concepts. For example, in Domencich and McFadden[13] the continuous variables producing discrete choices are differences in utilities of possible choices.

Only for convenience this threshold may be assumed to be zero. The random variable y_{it} is supposed to consist of two components: a deterministic component v_{it} which is a function of exogeneous, predetermined and measured endogeneous variables affecting current choices; and a purely random disturbance component ε_{it}

$$y_{it} = v_{it} + \varepsilon_{it} \tag{2.1}$$

with

$$y_{it} = \begin{cases} \geq 0 & \text{if } d_{it} = 1 \\ < 0 & \text{if } d_{it} = 0 \end{cases} \tag{2.2}$$

where d_{it} is a dummy variable denoting the occurrence of the event under consideration. The distribution of the d_{it}'s is generated by the distributions of the ε_{it} 's and v_{it} 's where adopting a multinomial probit formulation it is assumed that ε_{it} is normally distributed with mean zero and a (T,T)-positive definite covariance matrix. This normality assumption generates a model which admits a characterization of heterogeneity.

Assuming that the latent variable y_{it} is a linear function of observed choice-relevant attributes (including past exogeneous, current exogeneous variables and expectations of future exogeneous variables), represented in the vector $\boldsymbol{x}_{it}$, of lagged values y_{it} and of past outcomes $d_{it'}$ with $t' \leq t$. Heckman's general model may be written as:

$$v_{it} = \boldsymbol{x}_{it}\,\boldsymbol{\beta} + \sum_{j=1,\ldots,\infty} y_{i,t-j}\; d_{i,t-j} + \sum_{j=1,\ldots,\infty} \lambda_{jt-t} \prod_{l=1,\ldots,j} d_{it-l} + G(L)y_{it} \tag{2.3}$$

where $\boldsymbol{\beta}$ is a vector of parameters of $\boldsymbol{x}_{it}$; $G(0)=0$; and $G(L)=g_1 L+g_2 L^2 +\ldots$ is a general lag operator, with the property $L^k\, y_{it} = y_{it-k}$. The initial conditions $d_{it'}$ and $y_{it'}$ for $t' = 0,-1,-2,\ldots$ (in other words, the relevant presample history of the process) are assumed to be predetermined or exogeneous. This assumption, however, is only valid if the unobserved choice-relevant characteristics generating the process are serially independent.

The first term of the right hand side of (2.3) may incorporate past and current information and future expectations on exogeneous choice-relevant attributes affecting current choices, as already mentioned above. The second term represents the effects of the entire past history on the choice behaviour at time t and, thus, structural state dependence effects. The third term denotes the cumulative effect on current choices of the most recent experience in a state. The λ's denote parameters. Finally, the last term represents the effect of previous relative evaluations of the two states on current choices capturing the action of habit persistence.

Much progress has been made in panel data-based discrete time discrete choice modelling over the last few years. But unquestionably, there are several problems which have hitherto not been solved satisfactorily by these approaches. For example, the problem of initial conditions, the problem to account for heterogeneity due to variations outside the sample period and especially the problems related to correlations among individual decisions and the effect of non-stationarity. Hence, alternative approaches for the description of choice processes which avoid or at least circumvent difficulties inherent in the above discussed methods and enlarge the scope of possible applications of choice theory are highly welcome.

2.2 The Master Equation View in Dynamic Choice Processes

An interesting alternative to the panel data-based discrete time approach for analysing dynamic choice processes can be found in the so-called master equation approach. This approach which has a long tradition in physics (especially in the context of laser theory and spin relaxation) has been brought to the attention of the social sciences community in the late 1970s by Weidlich,[14] Weidlich and Haag,[15] and was then introduced in the regional sciences by Smith[16] and especially by Haag and Weidlich,[17-19] and Haag and Dendrinos.[20-21] In the last few years much research has been undertaken to open a large field of applications in the social and economic sciences in general and regional sciences in particular where special emphasis has been laid on the dynamics of migration processes.

A master equation describes the evolution of a probability distribution, representing the likelihood of a certain decision distribution of actors. The transition rates between well defined states of a dynamic micro-based system of actors influence the probability distribution. By using, for example, a mean value approach (see Section 2.4.3) an elegant link can be established between micro levels and macro levels of a system, so that structural changes in dynamic systems can be analysed in a statistically satisfactory way.

There are several cogent reasons for using the master equation approach in analysing dynamic choice processes. A first reason is its flexibility and generality. The ranges of possible behaviours embodied in master equations is almost unlimited (Smith[16]). Moreover, this approach enables researchers to take synergetic effects in the behaviour of different individuals (such as adaptation processes, learning effects, agglomeration effects) into account. A third major advantage of the master equation is that it links the microlevel decisions of individuals to the macrolevel behaviour of collective variables (see Fig. 2.2). The master equation offers a new approach to parameter estimation and feedback elements, heterogeneity (variation between individuals) and non-stationarity (variation over time) can be taken care of.

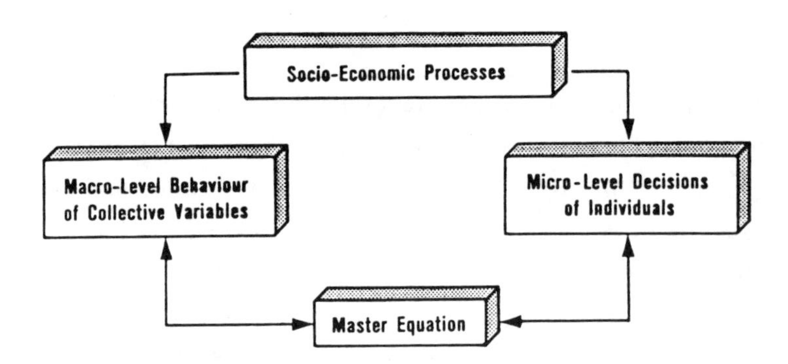

Figure 2.2: The master equation point of view: the relationship between the micro- and macrolevel in decision processes

The purpose of this section is to outline this approach in general terms and, moreover, to illustrate how the parameters involved in the choice model can be estimated. The relationship between the master equation and the static multinomial logit model will be outlined in the appendix (see Section 9.7). The construction of transition rates via panel data will be treated in Section 9.8.

Those readers who prefer to learn the method by means of examples may skip the rest of Section 2 and should proceed with the examples (Chapter 3 onwards).

2.3 The Decision Process

The central object of our dynamic theory of decision processes is to understand the decision behaviour of a group of *individuals* or *agents.* Interest therefore is confined to the observation of the time path of certain macrostates. Since the decisions have to be made under conditions of uncertainty, a stochastic description seems to be adequate. However, it is decisive to understand the link between the microlevel of decisions of single individuals or agents and the macrolevel defined by a few aggregate variables (mean values).

Of course, our theoretical concepts will have to include some simplifying, however, plausible assumptions. On the microlevel not all details having influence on the decision behaviour of an individual or agent are known or can be traced back to their origin. On the macrolevel on the other hand, we are in general restricted to only a few of empirically known data available for comparison with theory.

We proceed in three steps: first of all appropriate macrovariables of the socio-economic system have to be found. Second, we establish the equations of motion for these macrovariables. Third, we have to develop the methods necessary for applying our theory to concrete interactive decision processes.

2.3.1 Decision Space and Decision Configuration

We consider different kinds of individuals or agents in the system whose decisions control the whole socio-economic situation. Agents could be e.g., *consumers* whose decisions control *destination choice*, *households* who decide about their *place of residence*, *retailers* setting the *prices of goods or serv-*

ices, workers deciding about their *place of work, entrepreneurs* who decide about their *investment strategy, firms* making decisions about the *firm size* or the spectrum of produced *goods or services*. An agent could also be a group of individuals like a family or the managers of a firm.

It seems plausible to assume that the total population N of agents in the society consists of subpopulations P_α , $\alpha = 1, 2, \ldots, A$ (e.g., consumer, households, retailers, entrepreneurs) each with N_α members which are internally homogeneous but different from each other with respect to the decision behaviour of their members . Let $n_{\alpha i} \geq 0$ be the number of agents of subpopulation P_α having adopted state i of behaviour (having attitude i).

As an example we shall consider the migration of the native population ($\alpha = 1$) and the guest workers ($\alpha = 2$) in the Federal Republic of Germany. The migration behaviour of both population groups may differ considerably. Then n_{1i} , n_{2i} is the number of the native population and of the guest workers, living in region i ($i = 1, 2, \ldots, L$), respectively. The regions i are then the different federal states of the country.

The macrostate of the socio-economic system is now described by the *decision configuration*

$$\boldsymbol{n} = \{n_{\alpha i}\} \equiv \{n_{11}, \ldots, n_{A1}, \ldots, n_{\alpha i}, \ldots, n_{AL}\}, \tag{2.4}$$

consisting of the $C = AL$ integer variables $n_{\alpha i}$.

The N_α agents of the subpopulation P_α in general will have different attitudes i. Then

$$N_i = \sum_{\alpha=1}^{A} n_{\alpha i} \tag{2.5}$$

are all agents of the system of attitude i. The following relations are now obtained from the definition of the $n_{\alpha i}$.

$$N_\alpha = \sum_{i=1}^{L} n_{\alpha i} \tag{2.6}$$

and

$$N = \sum_{i=1}^{L} \sum_{\alpha=1}^{A} n_{\alpha i} . \tag{2.7}$$

The decisionconfiguration $\boldsymbol{n}$ describes the distribution of attitudes of the total population of agents at a given time.

However, it is sometimes useful to introduce a different notation for the word decision configuration in order to be more specific. For example, in the case of migration the notation *population configuration* is used or we speak of a *consumer configuration* if we consider consumers as agents.

Introducing the C - dimensional *decision space* $\mathbb{C}$, the time dependent decision distribution of the socio-economic system under consideration can be represented by a moving point $\boldsymbol{n}(t) \equiv \{n_{\alpha i}(t)\}$ in the decision space $\mathbb{C}$. This space $\mathbb{C}$ can be seen in analogy to the *phase space* for *macro variables* in statistical physics.

The individual decision processes of agents will now have to be related to the temporal variations of the decision configuration $\boldsymbol{n}(t)$.

2.3.2 Individual Decision Processes and Conditional Probability

The individual decisions of agents with respect to certain characteristics constitute the microlevel of consideration. However, a complete deterministic description of individual decisions is neither feasible nor desirable, since the details of motivations of the different agents are generally not available and on the other hand would overload the theory with too many details. We may circumvent this difficulty by introducing a probabilistic description.

The main purpose of this chapter is to demonstrate how a dynamic theory of decision processes can be established. The concept of the conventional static choice theory (see e.g., Domencich and McFadden,[13] Leonardi,[22] Wegener,[23] and Luce [24]) between alternatives cannot be used, since such a model of choice does not include the time factor at all (see also Section 9.7).

Instead we choose as a starting point the fundamental concept of the *conditional probability*

$$p(\boldsymbol{n}_0 + \boldsymbol{k}, t + \tau \mid \boldsymbol{n}_0, t; \text{past history}) \tag{2.8}$$

which is the probability of finding a decision configuration

$$\boldsymbol{n}_0 + \boldsymbol{k} \equiv \{n_{011} + k_{11}, \ldots, n_{0\alpha i} + k_{\alpha i}, \ldots, n_{0AL} + k_{AL}\} \tag{2.9}$$

at time $(t + \tau) \geq t$ with $\boldsymbol{k} \equiv \{k_{\alpha i}\}$, $k_{\alpha i} = 0, \pm 1, \pm 2. \pm 3, \ldots$, given that the initial configuration was

$$\boldsymbol{n}_0 \equiv \{n_{011}, \ldots, n_{0\alpha i}, \ldots, n_{0AL}\} \tag{2.10}$$

at time t. The expression (2.8) is zero by definition if $\boldsymbol{n}_0 + \boldsymbol{k}$ or $\boldsymbol{n}_0$ contain negative numbers $n_{0\alpha i} + k_{\alpha i}$ or $n_{0\alpha i}$, as these are invalid configurations. We shall now introduce some simplifying assumptions which decisively structure our further calculations:

a) Homogeneity assumption

We assume that the population of agents α is homogeneous with respect to their decision behaviour. This means that the conditional probability should be the same for all members of the subpopulation α

b) Markov assumption

Furthermore, the conditional probability should not depend on the decisions of the agents before time t. Hence, it is assumed that the decision-maker is unprejudiced by the previous history. He only takes into account the present decision configuration and a comparative socio-economic valuation of the consequences of all possible decisions at that particular time t

c) Assumption of statistical independence

We presume that the decisions of the *individual agents* contributing to the change $\mathbf{n}_0 \rightarrow \mathbf{n}_0 + \mathbf{k}$ are statistically independent of each other

d) Dependence of endogeneous and exogeneous variables

It is assumed that the conditional probability depends on endogeneous variables like the decision configuration itself and/or on exogeneous parameters called *trend parameters* $\boldsymbol{x}(t)$

As a consequence of the Markov assumption b) the conditional probability should not depend on its past history. This means that (2.8) assumes the form:

$$p(\mathbf{n}_0 + \mathbf{k}, t + \tau \mid \mathbf{n}_0, t) \tag{2.11}$$

The assumptions introduced above may not always be fulfilled in reality. Nevertheless, the effects of possible deviations from these assumptions will cancel out to a high degree on the macrolevel.

Since one of the decision configurations $\mathbf{n} = \mathbf{n}_0 + \mathbf{k}$ will always be attained at time $t + \tau$ when starting from any initial decision configuration $\mathbf{n}_0$, the probability normalization condition

$$\sum_{\{\mathbf{k}\}} p(\mathbf{n}_0 + \mathbf{k}, t + \tau \mid \mathbf{n}_0, t) = 1 \tag{2.12}$$

holds, where the sum $\{\mathbf{k}\}$ extends over all multiples of integers $\{k_{\alpha i}\}$ where $-\infty < k_{\alpha i} < +\infty$.

Furthermore, the initial condition must be fulfilled

$$p(\mathbf{n}_o + \mathbf{k}, t \mid \mathbf{n}_o, t) = \delta_{\mathbf{k},\mathbf{0}} \tag{2.13}$$

since the agents cannot change their opinion in the zero time interval $\tau = 0$. For very small time intervals τ the conditional probability (2.11) can be expanded into a Taylor series with respect to the variable $t' = t + \tau$ around t

$$p(\mathbf{n}_o + \mathbf{k}, t+\tau \mid \mathbf{n}_o, t) = \delta_{\mathbf{k},\mathbf{0}} + \tau\, w_t\, (\mathbf{n}_o + \mathbf{k}, \mathbf{n}_o) + O(\tau^2) \tag{2.14}$$

where

$$w_t\, (\mathbf{n}_o + \mathbf{k}, \mathbf{n}_o) = \left. \frac{\partial p(\mathbf{n}_o + \mathbf{k}, t' \mid \mathbf{n}_o, t)}{\partial t'} \right|_{t' = t} \tag{2.15}$$

Of course, for sufficiently small τ the higher order contributions can be neglected in (2.14). Inserting (2.13) into (2.14) yields for $\mathbf{k} \neq \mathbf{0}$

$$p(\mathbf{n}_o + \mathbf{k}, t + \tau \mid \mathbf{n}_o, t) = \tau\, w_t\, (\mathbf{n}_o + \mathbf{k}, \mathbf{n}_o) \tag{2.16}$$

while for $\mathbf{k} = \mathbf{0}$ it follows from (2.12) and (2.16), that

$$p(\mathbf{n}_o + \mathbf{k}, t + \tau \mid \mathbf{n}_o, t) = 1 - \sum_{\{\mathbf{k}\} \neq \{\mathbf{0}\}} \tau\, w_t\, (\mathbf{n}_o + \mathbf{k}, \mathbf{n}_o). \tag{2.17}$$

The new quantities $w_t\, (\mathbf{n}_o + \mathbf{k}, \mathbf{n}_o)$ are denoted as *configurational transition rates*. The $w_t\, (\mathbf{n}_o + \mathbf{k}, \mathbf{n}_o)$ describe the probability of change per unit time from a decision configuration $\mathbf{n}_o$ to a neighbouring decision configuration $\mathbf{n}_o + \mathbf{k}$.

2.3.3 Individual and Configurational Transition Rates

In this section we will consider how the configurational transition rates are related to transition rates on the level of individuals or agents. By definition, the *individual conditional probability* is the probability to find an individual decision maker - an agent - at a certain place in decision space at time $t + \tau$, given that he was initially, at time t, in a specific state in decision space $\mathbb{C}$. On the other hand, the *configurational conditional probability* is the probability to find a certain decision configuration $\boldsymbol{n} + \boldsymbol{k}$, at time $t + \tau$, given that the decision configuration $\boldsymbol{n}$ was realized at time t. Therefore, the relation between both kinds of conditional probabilities is a rather subtle combinatorial problem (Weidlich and Haag[25]).

Since, by assumption c) the agents make their decisions statistically independent of each other (but, the transition rates may depend on the decision distribution itself), the configurational probablility is given by the product of the individual conditional probabilities. For the shorttime propagator, namely the *configurational transition rate* this yields after some tedious but straightforward calculations the intuitively plausible relations presented below.

The configurational transition rate can be decomposed into different terms with respect to the nature of the underlying transition processes. Therefore, we start first of all by considering three completely different kinds of transition processes:

1. Agent changes his attitude

The transition processes in which one member of the subpopulation P_α of agents change from state (attitude) i to state (attitude) j of behaviour are very common. These individual transition rates per time unit are denoted by

$$p_{ji}^{\alpha}(\boldsymbol{n}, \boldsymbol{x}), \tag{2.18}$$

and depend in general on the decision configuration (socio-configuration) $\boldsymbol{n}$

and on trend parameters $\boldsymbol{x}$. The latter are considered as constants here, but may become dynamic variables in an extended treatment (see, for example Chapter 5 in Weidlich and Haag[15]).

As there exist $n_{\alpha i}$ members of agents of subpopulation P_α, the contribution of (2.18) to $w_t(\boldsymbol{n}+\boldsymbol{k}, \boldsymbol{n})$ is

$$w_{ji}^{\alpha}(\boldsymbol{n}+\boldsymbol{k}, \boldsymbol{n}) = n_{\alpha i}\, p_{ji}^{\alpha}(\boldsymbol{n}, \boldsymbol{x}), \tag{2.19}$$

for $\boldsymbol{k} = \{0, \ldots, 1_{\alpha j}, \ldots, 0, \ldots, -1_{\alpha i}, \ldots, 0, \ldots\}$, and

$$w_{ji}^{\alpha}(\boldsymbol{n}+\boldsymbol{k}, \boldsymbol{n}) = 0$$

for all other $\boldsymbol{k}$.

The vector $\boldsymbol{k}$ for non-vanishing transition rates $w_{ji}(\boldsymbol{n}+\boldsymbol{k}, \boldsymbol{n})$ contains zeros except for the integers 1 and (−1) in the positions αj, αi, respectively. The transitions (2.18, 19) conserve the numbers of agents N_α of the subgroup P_α.

2. Agent changes into another subpopulation

Another kind of transition occurs if an agent of subpopulation P_α changes per definition into another subpopulation P_β but keeps the previous attitude i. An example could be an individual's change from being single to married. The corresponding individual transition rate

$$p_i^{\beta\alpha}(\boldsymbol{n}, \boldsymbol{x}) \tag{2.20}$$

then contributes to $w_t(\boldsymbol{n}+\boldsymbol{k}, \boldsymbol{n})$ as follows:

$$w_i^{\beta\alpha}(\boldsymbol{n}+\boldsymbol{k}, \boldsymbol{n}) = n_{\alpha i}\;\; p_i^{\beta\alpha}(\boldsymbol{n}, \boldsymbol{x}) \tag{2.21}$$

for $\boldsymbol{k} = \{0, \ldots, 1_{\beta i}, \ldots, 0, \ldots, -1_{\alpha i}, \ldots, 0, \ldots\}$, and

$$w_i^{\beta\alpha}(\boldsymbol{n} + \boldsymbol{k}, \boldsymbol{n}) = 0$$

for all other $\boldsymbol{k}$.

The vector $\boldsymbol{k}$ for non-vanishing transition rates $w_i^{\beta\alpha}(\boldsymbol{n} + \boldsymbol{k}, \boldsymbol{n})$ contains zeros except for the integers 1 and (-1) in the positions βi and αi. The transition (2.20) conserves the number N_i .

3. Birth and death of agents

Finally birth and death processes are considered. The individual rates for the birth or death of an agent belonging to subpopulation P_α with attitude i are denoted by

$$p_{i+}^{\alpha}(\boldsymbol{n}, \boldsymbol{x}) \qquad \text{and} \qquad p_{i-}^{\alpha}(\boldsymbol{n}, \boldsymbol{x}) \tag{2.22}$$

and contribute to the configurational transition rate $w_t(\boldsymbol{n} + \boldsymbol{k}, \boldsymbol{n})$ as follows:

$$w_{i+}^{\alpha}(\boldsymbol{n} + \boldsymbol{k}, \boldsymbol{n}) = m_{\alpha i}\, p_{i+}^{\alpha}(\boldsymbol{n}, \boldsymbol{x}) \tag{2.23}$$

for $\boldsymbol{k} = \{\ldots, 0, \ldots, 1_{\alpha i}, \ldots, 0, \ldots\}$, and

$$w_{i+}^{\alpha}(\boldsymbol{n} + \boldsymbol{k}, \boldsymbol{n}) = 0$$

for all other $\boldsymbol{k}$ and

$$w_{i-}^{\alpha}(\boldsymbol{n} + \boldsymbol{k}, \boldsymbol{n}) = n_{\alpha i}\, p_{i-}^{\alpha}(\boldsymbol{n}, \boldsymbol{x}) \tag{2.24}$$

for $\boldsymbol{k} = \{\ldots, 0, \ldots, -1_{\alpha i}, \ldots, 0, \ldots\}$, and

$$w_{i-}^{\alpha}(\boldsymbol{n} + \boldsymbol{k}, \boldsymbol{n}) = 0$$

for all other $\boldsymbol{k}$. For birth processes $m_{\alpha i}$ in general can be assumed to be proportional to the number of agents of subpopulation P_{α}. The vector $\boldsymbol{k}$ for non-vanishing transition rates w_{i+}^{α}, w_{i-}^{α} contain zeros except for 1, (-1) in the respective positions αi.

It depends on the structure of the model being considered whether or not the (total) configurational transition rate $w_t(\boldsymbol{n} + \boldsymbol{k}, \boldsymbol{n})$ is the sum of *all* the contributions (2.19, 21, 23, 24) or whether some of them can be neglected. In the following chapters we shall discuss models from different fields. Therefore, the various contributions to $w_t(\boldsymbol{n} + \boldsymbol{k}, \boldsymbol{n})$ from individual transitions can be understood by means of examples.

In the general case the composition formula for the transition rate between decision configurations then reads:

$$\begin{aligned} w_t(\boldsymbol{n} + \boldsymbol{k}, \boldsymbol{n}) = & \sum_{j,i}\sum_{\alpha} w_{ji}^{\alpha}(\boldsymbol{n} + \boldsymbol{k}, \boldsymbol{n}) \\ & + \sum_{\alpha,\beta}\sum_{i} w_{i}^{\beta\alpha}(\boldsymbol{n} + \boldsymbol{k}, \boldsymbol{n}) \\ & + \sum_{\alpha,i} \{w_{i+}^{\alpha}(\boldsymbol{n} + \boldsymbol{k}, \boldsymbol{n}) + w_{i-}^{\alpha}(\boldsymbol{n} + \boldsymbol{k}, \boldsymbol{n})\}, \end{aligned} \tag{2.25}$$

where the index t indicates the possibility that the configurational transition rates themselves may still be time dependent. This may occur, for instance, when the trend parameters $\boldsymbol{\varkappa}$ become dynamic variables.

In (2.25) we have neglected contributions to the total configurational transition rate referring to simultaneous changes of an agent with respect to the subpopulation P_{α} he belongs to as well as his attitudes i. This means that a transition to a neighbouring state in configuration space $\boldsymbol{C}$ is assumed to be a sequential process.

2.3.4 Some Functional Forms of the Individual Transition Rates

In the previous sections we have stated that the crucial quantity of a dynamic decision process is the configurational conditional probability. For very short time intervals this quantity can be traced back to the configurational transition rates. However, the constitutive elements of the configurational transition rates are the individual transition rates.

The explicit construction of the individual transition rates has the purpose to attribute the information contained in the decision behaviour of certain individuals or agents to a *few* parameters only. The following is a very short list of possible suitable choices of the transition rates corresponding to the three cases mentioned in the last subsection.

case 1

Individual transition rate: agent changes his attitude

Let us first assume that two different states j, i in the decision space have the same advantage or disadvantage for an agent. From empirical observation we know that even in this case an agent could change his opinion (for instance from attitude i to attitude j) with a certain probability rate of transition. Therefore, we assign a trend parameter

$$\nu_{ji}^{\alpha} \geq 0 \tag{2.26}$$

denoted as *flexibility parameter* to this particular situation. The individual transition rate then reads:

$$p_{ji}^{\alpha}(\boldsymbol{n}, \boldsymbol{x}) = \nu_{ji}^{\alpha} \tag{2.27}$$

if the advantage of being in decision state j is equal to the advantage of being in state i.

However, it is reasonable to assume that in this particular situation the flexibility matrix (2.26) is symmetric:

$$\nu_{ji}^{\alpha} = \nu_{ij}^{\alpha} \tag{2.28}$$

Therefore, the number of open parameters is considerably reduced. It will be shown later, that in many applications the flexibility matrix can be reduced to a single parameter ν.

In the flexibility matrix (2.26) all effects should be included which will either facilitate or impede a transition from state i to j independent of any gain of advantage. Particularly *distance effects* in a general sense will influence the flexibility matrix.

We shall consider the concept of distance between i and j, if "state i" is interpreted as "living in region i". For this case *distance* is one of the most important specific variables in spatial analysis and many possibilities for its analytical specification are offered (Huriot and Thisse,[26] Kuiper[27]). However, the concept of *distance* must be generalized to comprise geographical as well as economic and social aspects. *Geographically* the length of routes between places i and j can often be used as an appropriate measure d_{ji} of distance. *Economic distances* are measured in terms of costs that can be evaluated in time or money. Traffic congestion can be important here. *Social distance* is connected with the incompleteness (diffusion) of information, concerning for example the vacant housing stock or the rent level distribution in the decision process of a potential mover. Vacant dwellings and job vacancies which are far away are less likely to be known to the potential mover who will therefore disregard them in his choice. It should be stressed that this discounting effect of distance on the knowledge about the choice set may in general be much more important than geographical or economic distance. The effect of geographical, economic and social distance is simultaneously measured by d_{ji}. In the following chapters we shall discuss some specific assumptions about the functional dependence:

$$\nu_{ji}^{\alpha} = \nu_{ji}^{\alpha}(d_{ji}). \tag{2.29}$$

On the other hand, all influences which are *asymmetric* with respect to the decision process are taken into account by introducing so called *dynamic advantage functions*

$$u_i^{\alpha}(\boldsymbol{n}, \boldsymbol{x}), \tag{2.30}$$

describing the *advantage* of an agent of subpopulation P_{α} to adopt state i of behaviour. Of course, this advantage will depend on the socio-economic situation of the system to be expressed by certain trend parameters $\boldsymbol{x}(t)$ and the configuration $\boldsymbol{n}(t)$. The different components of (2.30) can be regarded as *push/pull* terms, as will be discussed later.

There may exist relations of this concept of dynamic advantage functions and flexibility matrix to other conceptional frameworks like the static *random choice* theory. We shall discuss some of these relations later on. However, it must be stated, that the concept of dynamic advantage functions is a fully selfcontained and selfexplanatory one. The relevant quantities, namely the configurational transition rates $w_t(\boldsymbol{n} + \boldsymbol{k}, \boldsymbol{n})$ will be uniquely represented by the dynamic advantage functions and the flexibility matrix. This implies that their functional form implicitly defines the meaning and interpretation of the dynamic advantage function $u_i^{\alpha}(\boldsymbol{n}, \boldsymbol{x})$ and the flexibility matrix ν_{ji}^{α}. It should be pointed out that the concept of dynamic advatage functions as introduced here cannot be an *ordinal* one only taking into account the order (rank) of the advantage functions. It must be a *cardinal* concept in terms of explicit numerical values assigned to the $u_i^{\alpha}(\boldsymbol{n}, \boldsymbol{x})$.

Sometimes it seems to be intuitively more adequate to use other notations instead of the name dynamic advantage function. Depending on the field of application of the theory we will use the notation *utility function*, *profit function*, *welfare function*, and the like.

Let us now generalize the situation described by (2.27) and assume in a second step, that there exist for an agent of subpopulation P_α different possibilities in the decision space which are characterized by the same flexibility parameter ν^α_{ij} but distinguished by the magnitude of the corresponding dynamic advantage functions $u^\alpha_i(\boldsymbol{n}, \boldsymbol{x})$. Then the following minimal requirements must be fulfilled:

1. The individual transition rate p^α_{ji} is positive by definition. Therefore, p^α_{ji} must be a positive definite function of u^α_j and u^α_i for arbitrary real values of u^α_j and u_i

2. The individual transition rate p_{ji} from i to j must be larger than p_{ij} for the inverse transition, if u^α_j exceeds u^α_i

3. The individual transition rate p^α_{ji} must be a monotonously increasing function of the difference $(u^\alpha_j - u^\alpha_i)$, since an increasing advantage difference between j and i induces a higher probability to decide for j instead of i

The most reasonable form of $p^\alpha_{ji}(\boldsymbol{n}, \boldsymbol{x})$ satisfying these conditions is an exponential function

$$p^\alpha_{ji}(\boldsymbol{n}, \boldsymbol{x}) \sim \exp\{u^\alpha_j - u^\alpha_i\}. \tag{2.28}$$

Combining the flexibility part (2.27) and the advantage part (2.31) of the individual transition rate of an agent to decide for alternative j instead of i one is lead to the ansatz:

$$p^\alpha_{ji}(\boldsymbol{n}, \boldsymbol{x}) = \nu^\alpha_{ji}(d_{ji}) \exp\{u^\alpha_j(\boldsymbol{n}, \boldsymbol{x}) - u^\alpha_i(\boldsymbol{n}, \boldsymbol{x})\} \tag{2.32}$$

with

$$\nu_{ji}^{\alpha} = \nu_{ij}^{\alpha} . \tag{2.33}$$

The overall flexibility of subpopulation P_α with respect to changes in attitude is reflected in a time dependent flexibility parameter

$$\nu_0^{\alpha}(t) = \frac{1}{L(L-1)} \sum_{\substack{i,j \\ i \neq j}}^{L} \nu_{ji}^{\alpha}(t) \tag{2.34}$$

which is the mean value of the flexibility matrix.

Although (2.32) is highly plausible a few further remarks should be made to give additional reasons for the choice of this form of the individual transition rate:

4. We shall demonstrate in Section 9.7, that by using (2.32) the static random choice theory can be obtained as a stationary solution of our dynamic theory

5. Linearisation of the u_j^{α} with respect to socio-economic variables leads to nonlinear effects described by (2.32)

6. If the number of transitions per time unit from i to j of individuals or agents of subgroup P_α are empirically known (e.g., by counting), the trend parameters ν_{ji}^{α} and u_j^{α} can easily be estimated (see Section 2.3)

Since in (2.32) only the *differences* of the dynamic advantage functions of the alternative choice sets u_j^{α} and u_i^{α} appear, all u_j^{α} are only defined except for an arbitrary common additive constant. This constant can be adjusted in such a way that the dynamic advantage functions always fulfil the condition

$$\sum_{j=1}^{L} u_j^{\alpha}(\boldsymbol{n}, \boldsymbol{x}) = 0. \tag{2.35}$$

For further use we may introduce the variance of the dynamic advantage functions (making use of (2.35))

$$\sigma^2_{u\alpha} = \frac{1}{L} \sum_{j=1}^{L} (u_j^{\alpha})^2 \tag{2.36}$$

which is a measure for the inhomogeneity of the choice set with respect to the decision behaviour of the agents.

case 2

Individual transition rate: changes in subpopulation

A flexible formulation of the corresponding individual transition rate (2.20) of a change in subpopulation reads:

$$p_i^{\beta\alpha}(\boldsymbol{n}, \boldsymbol{x}) = \nu_i^{\beta\alpha}(\boldsymbol{n}, \boldsymbol{x})\exp\{u_i^{\beta\alpha}\} \tag{2.37}$$

where

$$\nu_i^{\beta\alpha}(\boldsymbol{n}, \boldsymbol{x}) = \nu_i^{\alpha\beta}(\boldsymbol{n}, \boldsymbol{x})$$

and

$$u_i^{\beta\alpha}(\boldsymbol{n}, \boldsymbol{x}) = - u_i^{\alpha\beta}(\boldsymbol{n}, \boldsymbol{x}). \tag{2.38}$$

The equations (2.37, 38) are generalizations of the ansatz (2.32, 33). In Chapter 4 an application using this more general formulation is given.

However, *Volterra-Lotka* processes also belong to this important field of individual transition rates (Goel and Richter-Dyn,[28] Getz,[29] Goel, Maitra and Montroll[30]). *Volterra-Lotka* processes or in other words *predator-prey* interactions are characterized by:

$$u_i^{\beta\alpha} = 0 \qquad \text{and} \qquad \nu_i^{\beta\alpha} = \mu_{\beta\alpha} n_{\beta i} \tag{2.39}$$

and yield the configurational transition rate:

$$w_i^{\beta\alpha} = \mu_{\beta\alpha} n_{\alpha i} n_{\beta i} .$$

case 3
Individual transition rate: birth/death processes

In a first approximation, the individual birth/death rates (2.22) or the specific birth/death rates can be considered as constants. However, density dependent effects (e.g., saturation effects) are well known and should be taken into account as well (Pielou,[31] Ludwig[32]). Plausible assumptions for (2.22) read:

$$p_{i+}^{\alpha}(\boldsymbol{n}, \boldsymbol{x}) = \nu_i^{\alpha}(1 - n_{i\alpha}/K_{i\alpha}) \qquad \text{for } n_{i\alpha} \le K_{i\alpha} \tag{2.40}$$

$$p_{i-}^{\alpha}(\boldsymbol{n}, \boldsymbol{x}) = \mu_i^{\alpha}(1 + n_{i\alpha}/M_{i\alpha}) \tag{2.41}$$

where $K_{i\alpha}$ is the *capacity* of the population P_α and $M_{i\alpha}$, and ν_i^α, μ_i^α are appropriate constants.

2.4 The Equations of Motion

In this section we shall derive equations of motion of the decision configuration. The description of the dynamics will be presented on both the *stochastic* and the *quasi-deterministic* level. However, only the *stochastic* or *proba-*

bilistic level is the fully exact and conceptually consistent one.

For only the fully probabilistic treatment gives insight into the way decisions on the microlevel of agents induce probabilistic fluctuations on the macrolevel. The equations of motion describing the full dynamics in probabilistic terms is the *master equation*. Thus the *master equation* provides the link between the microlevel and the macrolevel of consideration.

A short derivation of the master equation will be given in the next sub-section (Sect. 2.4.1). Because of the fundamental relevance of the master equation we give a more detailed derivation, embedded in the general concept of Markov-processes, in Chapter 9. Some important mathematical statements about the master equation as well as methods of its solution are also summarized in the appendix.

Although the method of the master equation is conceptually satisfactory and consistent, it has the disadvantage, that the master equation must be solved numerically in most cases and that its huge amount of information can only be compared with relatively poor empiric material.

In most cases, however, the *mean square deviations* from the *most probable path* of the system will be very small because of mutual cancellations of fluctuations. Then a *quasi-deterministic* (*mean value*) description is adequate. The approximate dynamic equations of the *mean values* can be directly derived from the master equation. This will be shown in Sect. 2.4.2.

2.4.1 The Master Equation for the Decision Configuration

Random decisions of individuals or agents constitute the basis of the dynamic evolution of the decision configuration (2.4). In the last section we have considered different processes causative for such changes in $\boldsymbol{n} = \{n_{\alpha i}\}$, on the microlevel. The probability to find a certain decision configuration $\boldsymbol{n}$ realized at time t, is denoted as the *configurational probability*

$$P(\boldsymbol{n}, t) = P(n_{11}, \ldots, n_{\alpha i}, \ldots, n_{AL}, t) . \tag{2.42}$$

Of course, $P(\mathbf{n}, t)$ must satisfy at all times the probability normalization condition

$$\sum_{\mathbf{n}} P(\mathbf{n}, t) = 1 \qquad (2.43)$$

where the sum extends over all possible decision configurations. If the configurational transition rates $w_t(\mathbf{n}+\mathbf{k}, \mathbf{n})$ from any $\mathbf{n}$ to all neighbouring $\mathbf{n}+\mathbf{k}$ are given, we can derive an equation of motion for $P(\mathbf{n}, t)$. This *master equation* is derived in detail in the appendix. It reads:

$$\frac{dP(\mathbf{n}, t)}{dt} = \sum_{\mathbf{k}} \{w_t(\mathbf{n}, \mathbf{n}+\mathbf{k})P(\mathbf{n}+\mathbf{k}, t) - w_t(\mathbf{n}+\mathbf{k}, \mathbf{n})P(\mathbf{n}, t)\}, \qquad (2.44)$$

where the sum on the right hand side of (2.44) extends over all $\mathbf{k}$ with non vanishing configurational transition rates $w_t(\mathbf{n}, \mathbf{n}+\mathbf{k})$ and $w_t(\mathbf{n}+\mathbf{k}, \mathbf{n})$, respectively. The master equation (2.44) has a very direct and intuitively appealing interpretation. The change with time of the probability of configuration $\mathbf{n}$ (left hand side of (2.44)) is due to two effects of opposite direction, firstly to the probability flux *from* all neighbouring configurations $\mathbf{n}+\mathbf{k}$ *into* $\mathbf{n}$ (first term of the right hand side) and secondly to the probability flux *from* $\mathbf{n}$ *to all* $\mathbf{n}+\mathbf{k}$ (second term of the right hand side).

The solution of this equation (2.44), namely the time dependent distribution $P(\mathbf{n}, t)$ contains all information about the choice process at the most detailed level. In particular not only the mean value of $\bar{\mathbf{n}}(t)$, but also their mean square deviations due to fluctuations in the decision process can be calculated. Correspondingly, the amount of mathematics to solve the time-dependent master equation is considerable, e.g., if there exist C different configurations $\mathbf{n} = \{n_{\alpha i}\}$, the master equations consist of C coupled linear differential equations for all $P(\mathbf{n}, t)$. The explicit form of this master equation will be written down in the following chapters for different socio-economic systems.

Inserting (2.25) in (2.44), the explicit form of the general master equation for dynamic decision processes can be obtained:

$$\frac{dP(\boldsymbol{n},t)}{dt} = \sum_{\boldsymbol{k}}\sum_{j,i}\sum_{\alpha}\{w_{ji}^{\alpha}(\boldsymbol{n},\boldsymbol{n}+\boldsymbol{k})P(\boldsymbol{n}+\boldsymbol{k},t) - w_{ij}^{\alpha}(\boldsymbol{n}+\boldsymbol{k},\boldsymbol{n})P(\boldsymbol{n},t)\}$$

$$+\sum_{\boldsymbol{k}}\sum_{i}\sum_{\alpha,\beta}\{w_{i}^{\beta\alpha}(\boldsymbol{n},\boldsymbol{n}+\boldsymbol{k})P(\boldsymbol{n}+\boldsymbol{k},t) - w_{i}^{\alpha\beta}(\boldsymbol{n}+\boldsymbol{k},\boldsymbol{n})P(\boldsymbol{n},t)\}$$

$$+\sum_{\boldsymbol{k}}\sum_{i}\sum_{\alpha}\{[w_{i+}^{\alpha}(\boldsymbol{n},\boldsymbol{n}+\boldsymbol{k})+w_{i-}^{\alpha}(\boldsymbol{n},\boldsymbol{n}+\boldsymbol{k})]\,P(\boldsymbol{n}+\boldsymbol{k},t)\}$$

$$-\sum_{\boldsymbol{k}}\sum_{i}\sum_{\alpha}\{[w_{i+}^{\alpha}(\boldsymbol{n}+\boldsymbol{k},\boldsymbol{n})+w_{i-}^{\alpha}(\boldsymbol{n}+\boldsymbol{k},\boldsymbol{n})]\,P(\boldsymbol{n},t)\} \tag{2.45}$$

The first, second and the last two terms on the right hand side of (2.45) describe probability flows in the decision space due to changes in the attitude space of agents, changes of agents into another subpopulation or birth/death processes, respectively.

2.4.2 The Translation Operator

In order to simplify the forthcoming derivations we introduce *translation operators*, $E_{\alpha i}$ acting on a function $f(\boldsymbol{n})$ of the decision configuration as follows:

$$E_{\alpha i}^{\pm 1} f(n_{11},\ldots,n_{\alpha i},\ldots,n_{AL}) = f(n_{11},\ldots(n_{\alpha i}\pm 1),\ldots,n_{AL}) \tag{2.46}$$

This definition implies the further relations:

$$n_{\gamma k} E_{\alpha i} E_{\alpha j}^{-1} f(\boldsymbol{n}) \equiv E_{\alpha i} E_{\alpha j}^{-1}(n_{\gamma k} - \delta_{\gamma\alpha}\delta_{ki} + \delta_{\gamma\alpha}\delta_{kj})\, f(\boldsymbol{n}) \tag{2.47}$$

$$n_{\gamma k} E_{\alpha i}^{\pm 1} f(\boldsymbol{n}) \equiv E_{\alpha i}^{\pm 1} (n_{\gamma k} \mp \delta_{\gamma\alpha} \delta_{ki}) f(\boldsymbol{n}) \tag{2.48}$$

and

$$E_{\alpha i} E_{\alpha j}^{-1} f(\boldsymbol{n}) \equiv f(n_{11}, \ldots, (n_{\alpha i} + 1), \ldots (n_{\alpha j} - 1), \ldots, n_{AL}). \tag{2.49}$$

Furthermore we introduce general configurations (GC):

$$\boldsymbol{n} = \{n_{11}, \ldots, n_{\alpha i}, \ldots, n_{AL}\}; \qquad n_{\alpha i} \gtrless 0 \tag{2.50}$$

whose components $n_{\alpha i}$ can be positive or negative integers or zero. The *true* decision configuration (DC) consists of such components only, which are composed of positive integers or zeros.

The functions $P(\boldsymbol{n}, t)$, $w_t(\boldsymbol{n} + \boldsymbol{k}, \boldsymbol{n})$ introduced so far are only defined for true configurations (DC). However, we can easily extend their definition to all general configurations (GC) by the requirement:

$$f(\boldsymbol{n}) = \begin{cases} f(\boldsymbol{n}) & \text{for } \boldsymbol{n} \in \text{DC} \\ 0 & \text{otherwise} \end{cases} \tag{2.51}$$

In all forthcoming calculations we make use of this formal extension. It implies that sums over $f(\boldsymbol{n})$ with $\boldsymbol{n}$ running through all true decision configurations, DC, can formally be extended to sums over all general configurations, GC.

It is evident, that then the relations hold:

$$\sum_{\boldsymbol{n}}^{GC} E_{\alpha i}^{\pm 1} f(\boldsymbol{n}) = \sum_{\boldsymbol{n}}^{GC} E_{\alpha i} E_{\alpha j}^{-1} f(\boldsymbol{n}) = \sum_{\boldsymbol{n}}^{GC} f(\boldsymbol{n}) \tag{2.52}$$

because the summation includes all contributions $-\infty \leq n_{\alpha i} \leq \infty$.

Inserting (2.19,21,23,24) in (2.45) and making use of the above introduced formalism, the master equation can be written in the equivalent form:

$$\frac{dP(\boldsymbol{n},\ t)}{dt} = \sum_{j,i}\sum_{\alpha}(E_{\alpha i}E_{\alpha j}^{-1} - 1)[\ n_{\alpha i}p_{ji}^{\alpha}(\boldsymbol{n},\ \boldsymbol{x})P(\boldsymbol{n},\ t)]$$

$$+ \sum_{i}\sum_{\alpha,\beta}(E_{\beta i}E_{\alpha i}^{-1} - 1)[n_{\alpha i}p_{i}^{\beta\alpha}(\boldsymbol{n},\ \boldsymbol{x})P(\boldsymbol{n},\ t)]$$

$$+ \sum_{i}\sum_{\alpha}(E_{\alpha i}^{-1} - 1)[m_{\alpha i}p_{i+}^{\alpha}(\boldsymbol{n},\ \boldsymbol{x})P(\boldsymbol{n},\ t)]$$

$$+ \sum_{i}\sum_{\alpha}(E_{\alpha i} + 1)[n_{\alpha i}p_{i-}^{\alpha}(\boldsymbol{n},\ \boldsymbol{x})P(\boldsymbol{n},\ t)] \tag{2.53}$$

This compact formulation of the master equation for decision processes will be the starting point for the derivation of appropriate mean value equations, in the next section.

2.4.3 The Mean Value Equations of the Dynamic Decision Theory

We mentioned that in general the probability distribution (configurational probability) $P(\boldsymbol{n},\ t)$ contains such a tremendous amount of information compared with the empirical situation that a less exhaustive description in terms of mean values seems to be adequate in most applications.

However, we expect that the mean value, $\bar{\boldsymbol{n}}\ (t)$, of the decision configuration $\boldsymbol{n}$ practically coincides with the *realized configuration.* Therefore, it is highly desirable to derive self-contained equations of motion for the mean decision behaviour of agents under given boundary conditions of a society.

We begin with the definition of the mean value $\overline{f(\boldsymbol{n})}$ of an arbitrary function $f(\boldsymbol{n})$:

$$\overline{f(\boldsymbol{n})} = \sum_{\boldsymbol{n}}^{DC} f(\boldsymbol{n})P(\boldsymbol{n}, t) = \sum_{\boldsymbol{n}}^{GC} f(\boldsymbol{n})P(\boldsymbol{n}, t) . \tag{2.54}$$

In particular the mean numbers of agents of subpopulation P_γ having adopted the state k of their choice sets are given by:

$$\overline{n}_{\gamma k}(t) = \sum_{\boldsymbol{n}}^{GC} n_{\gamma k} P(\boldsymbol{n}, t). \tag{2.55}$$

It is now possible to derive equations of motion for the mean values $\overline{n_{\gamma k}(t)}$ directly from the master equation (2.53). To this end (2.53) is multipied by $n_{\gamma k}$ and the sum over all configurations $\boldsymbol{n}$ is taken

$$\begin{aligned}
\frac{d\,\overline{n}_{\gamma k}(t)}{dt} &= \sum_{\boldsymbol{n}}^{GC} n_{\gamma k} \frac{dP(\boldsymbol{n}, t)}{dt} \\
&= \sum_{\boldsymbol{n}}^{GC}\sum_{j,i}\sum_{\alpha} n_{\gamma k}(E_{\alpha i}E_{\alpha j}^{-1} - 1)[\,n_{\alpha i}p_{ji}^{\alpha}(\boldsymbol{n}, \boldsymbol{x})P(\boldsymbol{n}, t)] \\
&+ \sum_{\boldsymbol{n}}^{GC}\sum_{i}\sum_{\alpha,\beta} n_{\gamma k}(E_{\beta i}E_{\alpha i}^{-1} - 1)[n_{\alpha i}\,p_{i}^{\beta\alpha}(\boldsymbol{n}, \boldsymbol{x})P(\boldsymbol{n}, t)] \\
&+ \sum_{\boldsymbol{n}}^{GC}\sum_{i}\sum_{\alpha} n_{\gamma k}(E_{\alpha i}^{-1} - 1)[m_{\alpha i}\,p_{i+}^{\alpha}(\boldsymbol{n}, \boldsymbol{x})P(\boldsymbol{n}, t)] \\
&+ \sum_{\boldsymbol{n}}^{GC}\sum_{i}\sum_{\alpha} n_{\gamma k}(E_{\alpha i} + 1)[n_{\alpha i}\,p_{i-}^{\alpha}(\boldsymbol{n}, \boldsymbol{x})P(\boldsymbol{n}, t)]
\end{aligned} \tag{2.56}$$

Making use of (2.46-53) and definition (2.54), on the right hand side of (2.56) we obtain the exact result:

$$\frac{d\,\overline{n_{\gamma k}}(t)}{d\,t} = \sum_{i \neq k}^{L} \overline{n_{\gamma i}\, p_{ki}^{\gamma}(\boldsymbol{n}, \boldsymbol{x})} - \sum_{j \neq k}^{L} \overline{n_{\gamma k}\, p_{jk}^{\gamma}(\boldsymbol{n}, \boldsymbol{x})}$$
$$+ \sum_{\alpha \neq \gamma}^{A} \overline{n_{\alpha k}\, p_{k}^{\gamma\alpha}(\boldsymbol{n}, \boldsymbol{x})} - \sum_{\beta \neq \gamma}^{A} \overline{n_{\gamma k}\, p_{k}^{\beta\gamma}(\boldsymbol{n}, \boldsymbol{x})}$$
$$+ \overline{m_{\gamma k}\, p_{k+}^{\gamma}(\boldsymbol{n}, \boldsymbol{x})} - \overline{n_{\gamma k}\, p_{k-}^{\gamma}(\boldsymbol{n}, \boldsymbol{x})} \quad . \tag{2.57}$$

The intuitive interpretation of (2.57) is obvious: the variation of $\overline{n_{\gamma k}(t)}$ with time is due to changes of agents of subpopulation P_{γ} with attitude i to attitude k, and vice versa, to changes of agents with attitude k of subpopulation P_{α} to subpopulation P_{γ}, and vice versa, and to birth/death processes of agents of subpopulation P_{γ} with attitude k.

The exact mean value equations still have the disadvantage to be not self-contained, because one needs the probability distribution $P(\boldsymbol{n}, t)$ in order to calculate the right hand side. If, however, it can be assumed that the probability distribution is a well behaved, sharply peaked unimodal distribution, the approximate relation /2.15/ holds

$$\overline{f(\boldsymbol{n}, t)} = f(\overline{\boldsymbol{n}}, t) \; . \tag{2.58}$$

The mean value of a function of $\boldsymbol{n}$ is assumed to be equal to that function of the mean value $\overline{\boldsymbol{n}}$.

Inserting the approximation (2.58) in (2.57), we obtain a closed set of self-contained equations of motion for the mean decision behaviour of agents

$$\frac{d\,\overline{n_{\gamma k}(t)}}{d\,t} = \sum_{i \neq k}^{L} \bar{n}_{\gamma i}\, p_{ki}^{\gamma}(\bar{\boldsymbol{n}}, \boldsymbol{x}) - \sum_{j \neq k}^{L} \bar{n}_{\gamma k}\, p_{jk}^{\gamma}(\bar{\boldsymbol{n}}, \boldsymbol{x})$$
$$+ \sum_{\alpha \neq \gamma}^{A} \bar{n}_{\alpha k}\, p_{k}^{\gamma\alpha}(\bar{\boldsymbol{n}}, \boldsymbol{x}) - \sum_{\beta \neq \gamma}^{A} \bar{n}_{\gamma k}\, p_{k}^{\beta\gamma}(\bar{\boldsymbol{n}}, \boldsymbol{x})$$
$$+ \bar{m}_{\gamma k}\, p_{k+}^{\gamma}(\bar{\boldsymbol{n}}, \boldsymbol{x}) - \bar{n}_{\gamma k}\, p_{k-}^{\gamma}(\bar{\boldsymbol{n}}, \boldsymbol{x}) \quad . \tag{2.59}$$

The equations (2.59) become fully explicit, by inserting the individual transition rates (2.32, 37, 40, 41). The mean value equations (2.59) establish a set of *A*L ordinary (nonlinear) coupled differential equations* for the $\overline{n_{\gamma k}(t)}$. Since the mean values by definition are *averages* over paths with fluctuating deviations, their evolution is described by *deterministic equations.* It must be expected, however, that the empiric values $n^{e}_{\gamma k}(t)$ show stochastic fluctuations around these mean values, even if the theory is adequate.

The mean value equations (2.59) belonging to the master equation (2.53) may have one or several stationary states. It will be shown, that they coincide with the maximum (the maxima) of the stationary decision distribution $P_{st}(\boldsymbol{n})$. All time dependent solutions approach for $t \to \infty$ one of these stationary states, but it depends on the initial conditions, which of the stationary states is approached (Eilenberger and Müller-Krumbhaar). This means, that our system of decision processes may - depending on the values of the trend parameters $\boldsymbol{x}$ - approach one unique, or one out of several possible, equilibrium state(s).

We shall now consider the case of trend parameters slowly varying with time $\boldsymbol{x} = \boldsymbol{x}(t)$. It is not important, whether this time dependence is considered as an exogeneous effect or as the result of an equation of motion for the trend parameters (endogeneous effect). Then it may occur, that the trend parameter passes a *critical value* for which one of the stationary equilibrium states of the system becomes unstable. However, if the system had beforehand adapted to this now unstable equilibrium, it will suddenly rush into a new stable equilibrium state or a new dynamic mode. We denote such a behaviour as a *phase transition*, in analogy to similar changes in physics. In the following chapters we will present different examples further illustrating this procedure.

After these general conclusions about the dynamics of decision processes we shall now discuss the comparison of our theory with empirically observed data.

2.5 Parameter Estimation

Our dynamic theory of decision processes, formulated in terms of a master equation or as a set of mean value equations is not yet operative, unless the main construction elements of these equations, the transition *rates* of agents with respect to their choice set, are specified. Since the latter depend on a set of trend parameters (e.g., flexibility parameters, dynamic utilities and the like), we must determine their values. Of course, it is well known, that it is difficult, if not impossible to establish a *direct and unique causal relationship* between the socio-economic situation and the decision behaviour of an agent or individual. Hence, we expect that many different influences merge into and will finally produce a certain decision pattern expressed by the transition rates. However, for some applications a rather good empiric data base is available. It then seems plausible to link the empiric socio-economic data to the theoretically assumed transition rates of the agents.

2.5.1 Parameter Estimation via Comparison of Transition Rates

In comparing the theoretical mean value equations (2.59) with the empirical decision behaviour of an agent, the mean configurational transition rate $w_t(\bar{n} + \boldsymbol{k}, \bar{n}; \boldsymbol{\varkappa})$ must be identified with the empirical transition rate corresponding to the respective transition (in this section we introduce explicitly the dependence of the transition rates on trend parameters $\boldsymbol{\varkappa}$):

$$w_t(\bar{n}+\boldsymbol{k}, \bar{n}; \boldsymbol{\varkappa}) \approx w_t^e(n^e+\boldsymbol{k}, n^e) \tag{2.60}$$

where

$$n^e(t) = \{n^e_{\alpha i}(t)\} \tag{2.61}$$

is the *empiric decision configuration* at time t, $t = 1, 2, \ldots, T$, and

$$w^e_t(n^e + k, n^e) \tag{2.62}$$

is the *empiric transition rate* from n^e to $n^e + k$.

By definition, the number of agents changing from the empiric configuration $n^e(t)$ to $n^e(t) + k$ in the time interval Δt is given by $\Delta t * w^e(n^e + k, n^e)$. The time interval Δt is given by e.g., the registration, counting or inquiry procedure and should be short enough to avoid multiple transitions.

The unknown large number of trend parameters x (e.g. v^α_{ij}, u^α_{ij},.) in (2.59) have to be estimated in an adequate fitting procedure. We apply the *method of least squares*, or in other words, we minimize the sum of squared deviations between the theoretical expression (2.25) and the empirical data (2.62)

$$F_t[x] = \sum_n \sum_k [w_t(n^e + k, n^e; x) - w^e_t(n^e + k, n^e)]^2 = \min. \tag{2.63}$$

Of course, other optimization procedures like the *maximum likelihood estimation* (Sen[34]) or a *nonlinear least square estimation* could also be used. In some numerical exercises, some of which are reported on in Sen and Pruthi,[35] the *least squares procedure* gave practically identical estimates to *maximum likelihood estimates.* Throughout this contribution we shall use the least square estimation since it is usually the simpler procedure.

For the numerical treatment of (2.63) the optimization procedures of Schwefel,[36] Schwefel and Drepper,[37] Drepper and Hermes,[38] Drepper et al,[39] Heckler[40] demonstrated their effectiveness in a high-dimensional parameter space.

It seems to be highly plausible to assume that the fluctuations of the $w_t^e(\boldsymbol{n}^e, \boldsymbol{m}^e)$ around their mean value are *Poisson - distributed* (Sen[41]). The assumptions a) - d) of Section 2.3.2 as well as empirical tests support this hypothesis. The minimization (2.63) can be expected to provide rather good fitting results for the trend parameters $\boldsymbol{x}(t)$. However, the *log-linear estimation* of the trend parameters $\boldsymbol{x}(t)$

$$G_t[\boldsymbol{x}] = \sum_{\boldsymbol{n}} \sum_{\boldsymbol{k}} [\log w_t(\boldsymbol{n}^e + \boldsymbol{k}, \boldsymbol{n}^e; \boldsymbol{x}) - \log w_t^e(\boldsymbol{n}^e + \boldsymbol{k}, \boldsymbol{n}^e)]^2 = \text{min.} \tag{2.64}$$

is also a frequently used procedure. It should be noted however, that going over to $\log w_t^e$ instead of w_t^e induces a bias which can be considerable for very small values of w_t^e. However, as shown in Sen and Pruthi[35] this bias is reduced quite substantially for moderately sized w_t^e, if we add 0.5 to w_t^e before taking the logarithm

$$\log w_t^e \rightarrow \log(w_t^e + 0.5). \tag{2.65}$$

The factor 0.5 in (2.65) is a consequence of the Poisson-assumption. The advantage of the log-linear estimation procedure (2.64) instead of (2.63) will become obvious in the next chapters. Nevertheless, a few words about their meaning should be mentioned here: because of the proposed exponential structure of the individual transition rates (2.32, 37) the log-linear estimation procedure yields quasi-linear relations between the trend parameters and empiric accessible data. Thus, a unique analytical expression for the estimated trend parameters can be derived. We recommend that both estimation methods (2.63, 64) be used and compared, in order to control the numerical output of (2.63) with the uniquely determined trend parameters according to (2.64).

2.5.2 Parameter Estimation via Comparison of Decision Configurations

In practice, however, the w_t^e are only available for exceptional cases. In most applications only stock variables like the $n_{\alpha i}^e(t)$ are accessible. Therefore, we have to look for other procedures which allow us to estimate the trend parameters $\boldsymbol{x}(t)$ even under such restrictive empiric conditions.

Then the following estimation procedure may yield satisfactory results: the time series of the trend parameters of our decision system are determined by the requirement that the appropriate mean value equations (2.59) describe the dynamics of the system as precisely as possible. Since a discrete set of empirical values $n_{\gamma k}^e(t)$ is used, the equation of motion yields a difference equation instead of a differential equation for the estimation procedure:

$$\begin{aligned}
\Delta n_{\gamma k}^e(t) &\approx n_{\gamma k}^e(t+1) - n_{\gamma k}^e(t) \\
&= \sum_{i \neq k}^{L} n_{\gamma i}^e \, p_{ki}^{\gamma}(\boldsymbol{n}^e, \boldsymbol{x}) - \sum_{j \neq k}^{L} n_{\gamma k}^e \, p_{jk}^{\gamma}(\boldsymbol{n}^e, \boldsymbol{x}) \\
&\quad + \sum_{\alpha \neq \gamma}^{A} n_{\alpha k}^e \, p_k^{\gamma\alpha}(\boldsymbol{n}^e, \boldsymbol{x}) - \sum_{\beta \neq \gamma}^{A} n_{\gamma k}^e \, p_k^{\beta\gamma}(\boldsymbol{n}^e, \boldsymbol{x}) \\
&\quad + m_{\gamma k}^e \, p_{k+}^{\gamma}(\boldsymbol{n}^e, \boldsymbol{x}) - n_{\gamma k}^e \, p_{k-}^{\gamma}(\boldsymbol{n}^e, \boldsymbol{x})
\end{aligned}$$

$$\text{for } t = 1, 2, \ldots, T. \qquad (2.66)$$

Assuming that the trend parameters $\boldsymbol{x}(t)$ vary with time only slowly, the optimal set of the parameters can be found by the requirement that the functional

$$F_t[\boldsymbol{x}(t)]$$

$$= \sum_{t'=t-\tau}^{t+\tau} \sum_{\gamma,k} \Big\{ \Delta n^e_{\gamma k}(t') - \Big[\sum_{i \neq k}^{L} n^e_{\gamma i}\, p^{\gamma}_{ki}(\boldsymbol{n}^e, \boldsymbol{x}) - \sum_{j \neq k}^{L} n^e_{\gamma k}\, p^{\gamma}_{jk}(\boldsymbol{n}^e, \boldsymbol{x})$$

$$+ \sum_{\alpha \neq \gamma}^{A} n^e_{\alpha k}\, p^{\gamma\alpha}_{k}(\boldsymbol{n}^e, \boldsymbol{x}) - \sum_{\beta \neq \gamma}^{A} n^e_{\gamma k}\, p^{\beta\gamma}_{k}(\boldsymbol{n}^e, \boldsymbol{x})$$

$$+ m^e_{\gamma k}\, p^{\gamma}_{k+}(\boldsymbol{n}^e, \boldsymbol{x}) - n^e_{\gamma k}\, p^{\gamma}_{k-}(\boldsymbol{n}^e, \boldsymbol{x}) \Big] \Big\}^2 = \text{min.} \tag{2.67}$$

be minimized in the interval $t' = [t - \tau, t + \tau]$. In (2.67) we introduced a moving average with step size $(2\tau + 1)$, in order to smooth irregular effects caused by fluctuations in the empirical data. The determination of the trend parameters by solving (2.67) constitutes a complex nonlinear optimization problem. Especially evolutionary strategies with self-learning algorithms seem to be most efficient to overcome stagnation in the numerical processing. The overwhelming success of mixed strategies is shown by Schwefel.[42] In order to reduce CPU-time, parallel-processor machines or multi-processor machines are highly recommended.

In the next section we give a short account of how trend parameters can be regressed to motivation factors.

2.5.3 The Dependence of Trend Parameters on Motivating Factors

We now consider the case in which empirical analysis exhibits a pronounced time dependence of the trend parameters. It may then be promising to analyse the correlation between trend parameters $\boldsymbol{x}(t)$ and possible motivation factors $\boldsymbol{\Omega}(t) = \{\Omega_1(t),\dots, \Omega_S(t)\}$, creating the spatio-temporal variations. Typical time and space dependent variables available are for instance: income per capita, labour force, housing stock, rents, taxes. We assume that the socio-economic motivation factors are properly standardized and detrended, in the time interval $t = 1, 2, \ldots, T$ under consideration.

Although it cannot be contested, that an influence of such socio-economic variables on decision making exists, it is rather difficult if not impossible to establish such a relationship directly. Whatever the causal relationship between the $\boldsymbol{\Omega}(t)$ and the decision process may be, it is justified to assume a tentative relation between the trend parameters $\boldsymbol{\varkappa}(t)$ and the motivation factors $\boldsymbol{\Omega}(t)$

$$\varkappa_r(t) = \sum_{s=1}^{S} a_{rs}\, \Omega_s(t - \tau_{rs}) \qquad \text{for } t = 1, 2, \ldots, T \quad \text{and } r = 1, 2, \ldots, S \tag{2.68}$$

where we have introduced eventual time-lags τ_{rs}.

The degree of influence of socio-economic variables on the trend parameters of the decision process will vary considerably. Therefore, algorithms for the ranking of relevance of socio-economic variables are highly welcome (see Reiner and Munz[43]). It is the purpose of this algorithm[43] to select objectively a few but relevant socio-economic variables which then will be denoted as *key-factors*.

The following chapters illustrate our dynamic theory of decision processes by examples. The different examples are selected in order to stress different important aspects of model building in the broad field of socio-economic processes. In all cases, however, the decision processes of agents are the central part and the starting point of the models considered.

2.5.4 Scheme of Model Building for Dynamic Decision Processes

The constitutive elements of our dynamic theory of decision processes are summerized in Fig. 2.3. In order to facilitate their identification the conceptional steps and the logical connections are indicated by arrows.

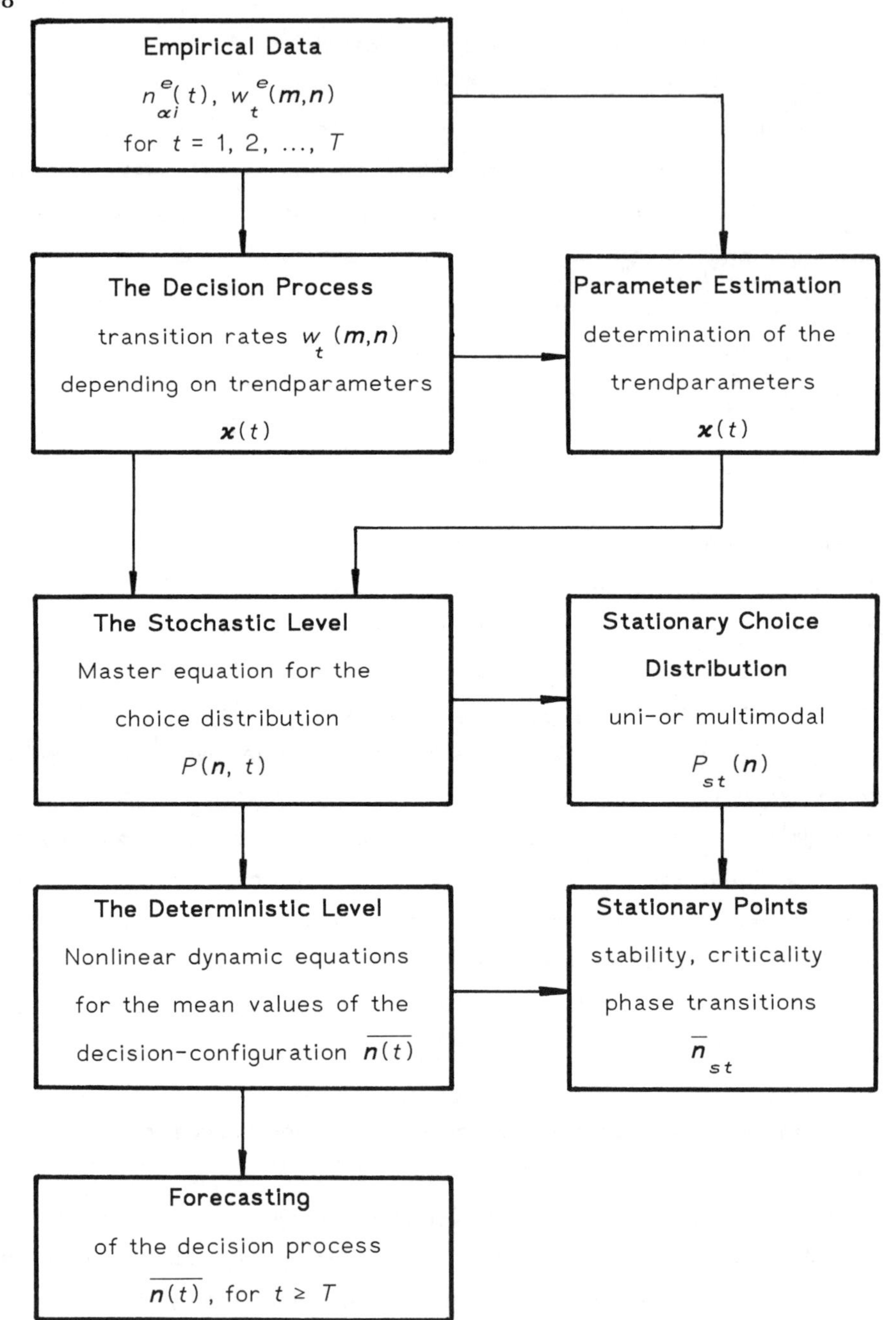

Figure 2.3 : Scheme of model building

2.6 Selection Criteria for the Examples

In this chapter the important tools of this dynamic theory of decision processes has been introduced. Together with some general concepts of probability theory, the derivation of the master equation from first principles, and an assortment of a number of important properties of the master equation in the appendix, the theoretical part of this book has been completed.

In the following, the ideas and concepts will be illustrated by means of examples with increasing complexity in each chapter. The regional sciences provide an interesting and wide field of potential applications of this theory mainly because of the following reasons:

- **Huge variety of topics**

  A huge variety of problems in the regional sciences can be formulated as a choice process. The individual agents and their choice strategy is relatively well defined (at least in migration theory)

- **Synergetic interactions among individuals**

  The interactions among the different agents (consumers, households, developers, retailers, firms, etc.,) lead to an interesting complexity and require sophisticated methods. Synergetic effects (nonlinearities) are necessary to understand the spatio-temporal pattern of urban systems

- **Longitudinal survey data**

  Panel data provide the information on a (more or less) fixed sample of decision-makers during time so that statements can be made about the behavioural response at the individual level. Dynamic models of discrete choice have to be based on panel data

- **Micro/macro relationship**

  One of the most important questions in social sciences is the micro-macro relationship. The chosen examples allow the researcher to consider this fundamental point in more detail

- **Non-stationarity**

  It is empirically evident, that the dynamic aspect plays a more and more fundamental role in all spatio-economic systems. In comparison with equilibrium models we expect better results and deeper insights via a dynamic approach

- **Scientific importance**

  Urban growth and decline has a strong influence on all spheres of society. Hence, the political and social consequences are of crucial importance to all individuals

In order to avoid an overloaded and disordered second part of this book, all examples are taken from the field of regional sciences. It is obvious that especially in economics a huge field of further applications can be found. Moreover, the above mentioned reasons together with the wish to provide a selfconsistent contribution make it advisable to restrict ourselves and to choose all examples from the field of regional sciences. However, the complexity of the different examples increases in order to model more relevant details gradually.

Chapter 3

Shocks in Urban Evolution

Why were there only five cities of over one million inhabitants (Berlin, London, Paris, Peking, and Vienna) in the world in 1860, and nowadays well over two hundred ? Finding the answer to this question poses one of the greatest challenges to regional scientists. This tremendous growth of settlements influences the structure of society in the developed as well as in the less developed world (Figure 3.1) Nevertheless, the political and social consequences of shocks in urban growth or decline are not sufficiently taken into consideration and there exist as yet no adequate strategies or recommendations for planners.

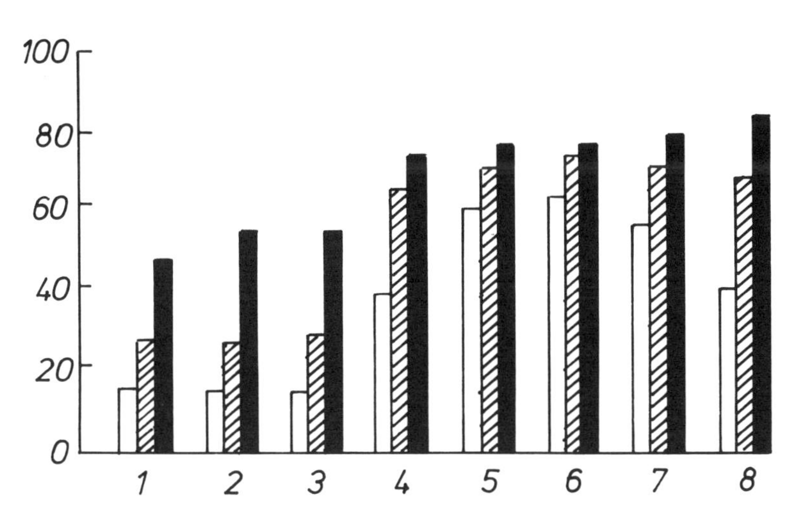

Figure 3.1: Urbanization trends ▭ 1950, ▨ 1985, ■ 2025 (UN predictions)
1: East Asia, 2: South Asia, 3: Africa, 4: USSR, 5: Australasia, 6: North America, 7: Europe, 8: Latin America.
Source: *The Economist*, 1987

3.1 Introduction

Since the growth rate of the total world population is lower than the rate of the cities population growth, sudden urban growth must feed on migration. This suggests that the migrant sees a clear advantage of living in a city over all other alternatives (see Papageorgiou [1]). Furthermore the facts tell us that this attractiveness of the city leads to its sudden growth at a certain threshold of size followed by a more continuous growth afterwards (see Figure 3.2).

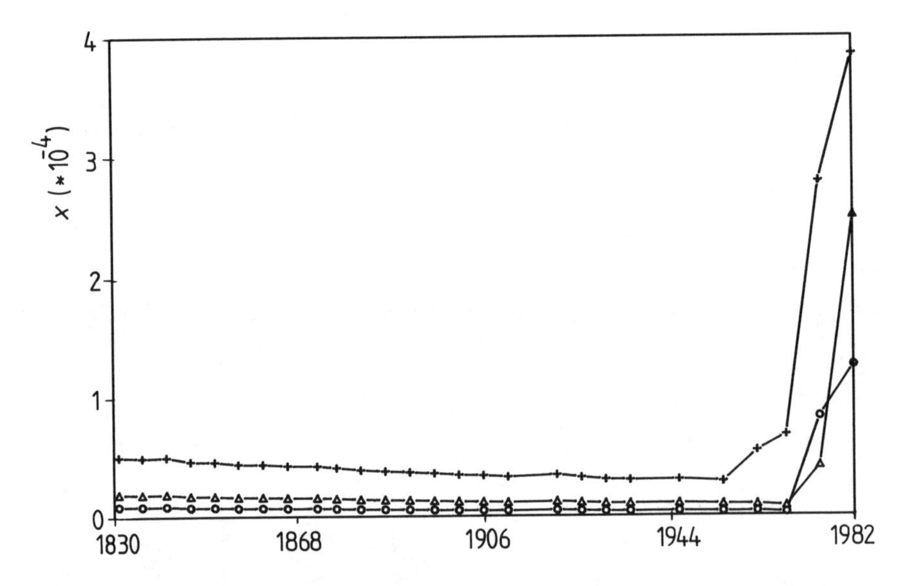

Figure 3.2: The relative population share of different cities of the French urban system. + Roissy, o Othis, Δ Villefontaine

In order to explain this phenomenon Wheaton[2] introduced, on the basis of microeconomic considerations a relationship between the urban opportunities and the size of a city

$$u = u(n_c) \tag{3.1}$$

where $u(n_c)$ is the utility for an individual to stay in the city, and n_c is the population size of the city.

In order to examine the consequences of (3.1) let us consider a system consisting of a *city* with population size n_c and its *hinterland* with population size n_h. It can be assumed that the total population size $n_c + n_h$ remains constant and is partitioned between these two areas (see Figure 3.3). Furthermore, it is reasonable to assume (Casetti[3]) that

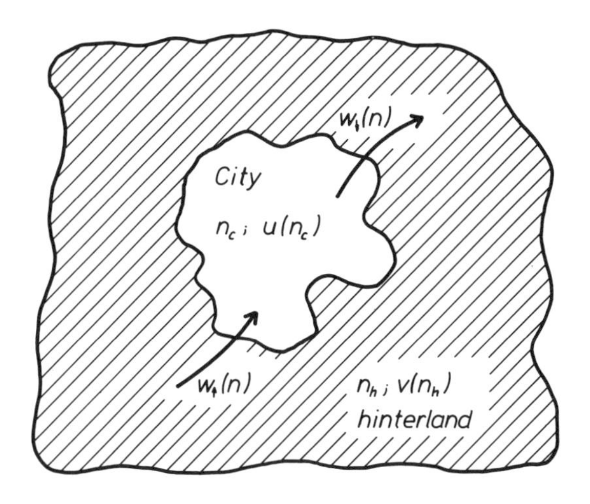

Figure 3.3: The city-hinterland system and its migratory variables

the utility v of the alternative area (the hinterland) also depends on its population size n_h

$$v = v(n_h). \tag{3.2}$$

According to *conventional* microeconomic principles, each individual tries to maximize his utility under certain given constraints (e.g., budget constraints). Therefore, if the urban utility exceeds the utility of the hinterland

$$u(n_c) > v(n_h), \tag{3.3}$$

with everything else remaining unchanged, a migration towards the city is caused. From the above mentioned microeconomic point of view, the equilibrium is reached if the population of the system is split in such a way that for each individual an equal utility level is obtained

$$u(n_c^{eq}) = v(n_h^{eq}). \tag{3.4}$$

Obviously, under this condition neither an individual in the hinterland nor one in the city can gain a utility advantage by changing his residence from h to c or vice versa. That implies that migration stops completely.

We shall now consider utility functions $u(n_c)$, $v(n_h)$, which only depend linearly on population size

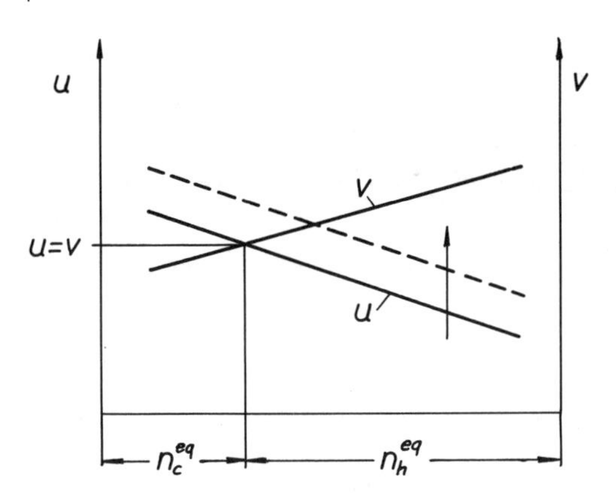

Figure 3.4: Utility variation with city size according to (3.5). The utilities $u(n_c)$, $v(n_h)$ may shift upwards or downwards, e.g., due to technological development

$$u(n_c) = \alpha_0 + \alpha_1 n_c$$

$$v(n_h) = \beta_0 + \beta_1 n_h \tag{3.5}$$

If it is assumed, as in Wheaton,[2] that an increase of population has *negative* effects on the levels of the utilities, or in other words, *if*

$$\alpha_1 < 0 \qquad \text{and} \qquad \beta_1 < 0. \tag{3.6}$$

we find a situation as plotted in Figure 3.4.

In this figure, the intersection between $u(n_c)$ and $v(n_h)$ marks the equilibrium partition of population between the city and the alternative sector (hinterland) according to (3.4). The development of technology will in the course of time shift the level of the utilities u, v, upwards.

Assuming that the city is more responsive to such technological changes than the hinterland, a continuously growing equilibrium size of the city may thus be obtained. However, it is obvious from Figure 3.4, that under such conditions a sudden growth of the urban size cannot be explained.

The plausibility of the assumption, that the utilities of $u(n_c)$ and $v(n_h)$ are *decreasing* with n_c and n_h, respectively, may also be doubted with good reasons.

In order to obtain *sudden* (instead of smooth) urban growth under the equilibrium condition (3.4) Papageorgiou proposed a third-order polynomial for the urban utility function with respect to the cities population size n_c

$$u(n_c) = \alpha_0 + \alpha_1 n_c + \alpha_2 n_c^2 + \alpha_3 n_c^3 \tag{3.7}$$

The coefficients α_i and their signs are dependent on economic variables such as e.g., the marginal productivity of labour used in the production process. While Wheaton failed to capture the relationship between urban productivity and urban size and therefore excludes returns to scale in agglomeration, Papageorgiou introduced urban production explicitly.

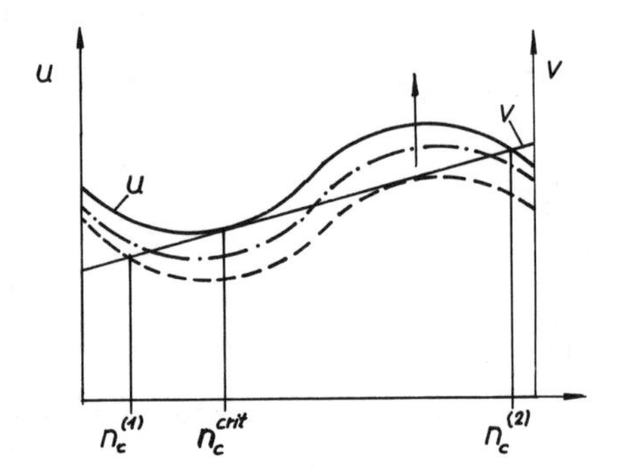

Figure 3.5: Utility variation with city size according to (3.7). The utility $u(n_c)$ shifts upwards with development of technology

In Figure 3.5 a specific form of the utility functions (3.7), $u(n_c)$ and $v(n_h)$ is depicted. Shifts in the absolute level of the utilities now cause smooth changes of the population size of a city until the instability point n_c^{crit} is reached. A small further upward shift of the utility $u(n_c)$ then leads to a dramatic sudden transition from n_c^{crit} to $n_c^{(2)}$ of the equilibrium city size.

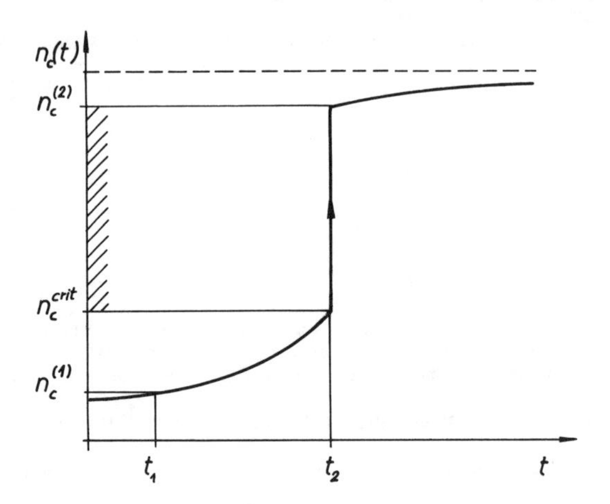

Figure 3.6: Equilibrium urban population growth according to Figure 3.5 (range of bistability /////)

In Figure 3.6 the corresponding schematic trajectory of the equilibrium sizes of a city is shown. Obviously, the discontinuity of the depicted trajectory relates to the transition of the stable intersection point from n_c^{crit} to $n_c^{(2)}$.

However, some critical remarks should now be made about models starting from the equilibrium condition (3.4).

The situation characterized by (3.4) and by zero migration in both directions turns out to be too restrictive and also not to be realized. From the macroscopic standpoint a better definition of equilibrium between the city and the hinterland is given by the requirement of *equal population flows* (that means zero net migration) between the two regions. Also from empirical observations it seems more reasonable to assume that individuals keep a certain mobility, even if no utility gain is connected with a move (that means, even if (3.4) is fulfilled). Under the condition of equal individual mobility from h to c and c to h, however, an imbalance in population sizes, e.g., $n_h > n_c$, will lead to an excess of the migration flow from h to c against the counter flow from c to h. Because of (3.1) and (3.2), this net-migration in turn will create an imbalance between the utilities $u(n_h)$ and $u(n_c)$ in such a way that the net-migration decreases. The macroeconomic equilibrium is eventually reached for *zero net migration* and will be connected with a stationary utility difference between the city and its hinterland (in contrast to (3.4)). We shall come back to this question in Section 3.2.4. (An interesting analogy to this consideration out of Physics can be seen in the urn experiment of Ehrenfest[4])

Furthermore, it is questionable and seems rather artificial to assume that individuals react in such a sensitive and nonlinear way on density variations of a certain region as required by the form (3.7) of the utility function. Of course, it could be argued that people are sensitive to income, availability of land or other socio-economic variables and that the nonlinearity of utility function $u(n_c)$ is merely the result of the complicated reactions of individuals to existing socio-economic imbalances. But then an explicit justification of form (3.7) of the utility function is urgently needed.

The question therefore must be raised whether less complicated assumptions about $u(n_c)$ can yield an explanation of the phenomenon of sudden urban growth if simultaneously the equilibrium condition (3.4) is substituted by the more realistic stationary condition of vanishing net flow between c and h.

It is this alternative approach which will be worked out in detail in the next subsection (see Haag[5]).

3.2 A Stochastic Model on Shocks in Urban Evolution

Starting point of our considerations is the empirical fact, that the take off of cities cannot be explained by simple birth-death processes only. Thus, sudden urban growth feeds on migration.

We consider a city, denoted by c and an alternative sector h (hinterland) according to Fig. 3.3. The number of people living in the city is n_c , and the number of people living in the alternative sector is n_h.

For notational purposes the following variables are introduced:

$$n_c + n_h = 2N \qquad \text{and} \quad n_c - n_h = 2n$$
$$n_c = N + n \qquad \text{and} \quad n_h = N - n \tag{3.8}$$

where

$$-N \leq n \leq N \qquad \text{and} \quad 0 \leq n_c, n_h \leq 2N$$

N denotes half of the total population size (for mathematical convenience only). Neglecting birth and death processes in a first approach, we can consider N as a constant, while n remains *the only relevant variable*. The effect of birth and death rates leads to variations in the total population size and will be discussed later.

In general not all residents of the city and the hinterland are potential participants of the migration dynamics. Instead there is an *inert part* and a *mover pool part* of the population in both regions. The numbers n_c , n_h and N

then refer to the mover pool populations only.

The problem is further simplified by assuming a homogeneous population whose members have the same individual decision rates with respect to their migratory behaviour.

The population configuration $\boldsymbol{n} = \{ n_c, n_h \}$ then consists of the numbers n_c and n_h of individuals having residence in the city c or or the hinterland h, respectively. However, as already mentioned, it suffices to characterize the configuration $\{n_c, n_h\}$ by one variable only, namely by $n = N - n_c = n_h - N$. The population configuration is not only of theoretical interest, but is also empirically known, at least in principle, by the repeated registration at times $t = 1, 2, \ldots, T$ of the number of people in each region. We denote the empiric population configuration at time t by the index e:

$$\boldsymbol{n}^e(t) = \{n_c^e(t), n_h^e(t)\} \qquad \text{for } t = 1, 2, \ldots, T \tag{3.9}$$

The other set of empirically known quantities are the components of the migration matrix $w_{\uparrow}^e(t)$, $w_{\downarrow}^e(t)$. The $w_{\uparrow}^e(t)$, $w_{\downarrow}^e(t)$ describe the number of migrants per time unit from hinterland to the city and from the city to the hinterland, respectively. The time interval (time unit) is given by the registration procedure. It should be short enough so that practically nobody migrates twice per time unit.

3.2.1 The Configurational Transition Rates

Let us now consider the microlevel of migration. It consists of individuals' decisions to migrate from the city to the hinterland, and vice versa.

Therefore, we consider as *agents* single individuals or more generally a group of individuals (family) with a certain mean size. However, a completely deterministic description of individual decisions is neither feasible nor desirable, since the details of individual motivations to change location are not available and would overload the theory with too many details. The way out of this difficulty is the transition to a probabilistic description as pointed out in Chapter 2.

In this particular case the individual decision rates of the agents induce "nearest neighbour" transitions of the population configuration (see Sect. 2.1.3) only. The transition

$$\{n_c, n_h\} \rightarrow \{n_c + 1, n_h - 1\}, \quad \text{or equivalently} \quad n \rightarrow n + 1$$

is effected by the migration of one of the n_h members of the population of the hinterland h to the city c. Analogously, the transition

$$\{n_c, n_h\} \rightarrow \{n_c - 1, n_h + 1\}, \quad \text{or equivalently} \quad n \rightarrow n - 1$$

is effected by the migration of one of the n_c population members of the city c to the hinterland h. Correspondingly, the *configurational transition rates* for the whole population distribution are given by

$$w((n+1) \leftarrow n) \equiv w_\uparrow(n) = n_h\, p_\uparrow(n) = (N - n) p_\uparrow(n)$$
$$w((n-1) \leftarrow n) \equiv w_\downarrow(n) = n_c\, p_\downarrow(n) = (N + n) p_\downarrow(n)$$

and

$$w(n' \leftarrow n) = 0 \qquad \text{for} \quad n' \neq n \pm 1. \tag{3.10}$$

where the $p_\uparrow(n)$, $p_\downarrow(n)$ are the *individual transition rates* (individual decision rates) of an individual agent to opt for his place of residence in the city instead of the hinterland, or vice versa. The $w_\uparrow(n)$, $w_\downarrow(n)$ are the corresponding *configurational transition rates* for a change of the whole *population configuration*.

3.2.2 The Master Equation

We shall now develop equations of motion for the dynamics of the population configuration. Since the individual decision process is described in probabilistic terms, the evolution on the macrolevel can only be probabilistic, too. Therefore, only the fully probabilistic treatment gives an insight into the way in which the decisions on the microlevel of individuals induce probabilistic fluctuations on the macrolevel. The mean square deviations on the macrolevel from the deterministic path can, however, be very small because of mutual cancellations of fluctuations. The equation of motion for the probability that a certain population distribution is realized is the master equation (Chapters 2, 9). Therefore, the master equation provides the link between the microlevel of individual decision processes and the macroeconomic equations of motion.

The underlying process of the evolution of the population configuration consists of the decisions of individuals to migrate. Of course, these decisions bear some uncertainty. The probabilistic description of these decision processes was given in Section 3.2.1. Therefore, on the macrolevel in the strict sense only probabilistic statements are possible. The fundamental quantity is the probability to find realized at a given time t a certain population configuration $\{n_c, n_h\}$. This configurational probability is denoted as

$$P(n, t) = P[n_c, n_h, t]. \tag{3.11}$$

Of course, the configurational probability must satisfy at all times t the normalization condition

$$\sum_{n=-N}^{N} P(n, t) = 1 \tag{3.12}$$

where the sum extends over all possible population configurations.

The equation of motion to be derived for $P(n, t)$ is the master equation (for a general derivation see Chapter 9). In our particular example the master equation reads:

$$\frac{d\,P(n,t)}{dt} = w_{\downarrow}(n+1)P(n+1,t) + w_{\uparrow}(n-1)P(n-1,t) - (w_{\downarrow}(n)+w_{\uparrow}(n))P(n,t)$$

$$\text{for } -N \leq n \leq N \qquad (3.13)$$

This equation has a very direct and intuitively appealing interpretation: The left hand side of (3.13) is the change per time unit of the probability of a given population distribution n. The right hand side of this equation describes the probability flow from neighbouring configurations ($n \pm 1$) to the configuration n and the probability flow from configuration n to all neighbouring configurations ($n \pm 1$). Summarizing we can state that the master equation (3.13) is nothing but a probability rate equation describing the change of the probability of every configuration in terms of the net probability flow between n and all neighbouring configurations n'.

3.2.3 The Individual Transition Rates

By insertion of the configurational transition rates (3.10) into the master equation (3.13) our problem is traced back to the important question, how the migratory trends can optimally be reflected in the functional form of the individual transition rates $p_{\uparrow}$, $p_{\downarrow}$. It is the aim to attribute the information contained in the migration flows to a few but meaningful parameters only, having well discernible independent interpretations. In principle two procedures appear to be possible:

a) Direct empirical determination of the $p_{\uparrow}$, $p_{\downarrow}$ within an ensemble of individuals moving between h and c. This direct method is relatively difficult to apply, since the decision finding process is rather complex and could be biased by inquiry

b) Plausible assumptions on the functional dependence of the transition rates can be developed. Inserting the specific form developed into the equations of the model results in certain structures for the population configuration. These structures could then be compared with reality in order to verify or falsify the plausibility of the chosen transition probabilities

In this Section we shall restrict ourselves to the second alternative. The individual transition probabilities $p_{\uparrow}$, $p_{\downarrow}$ are assumed to be functions of the momentary population configuration n. We assume further, that the potential migrant compares the utilities of the origin region and the prospective destination region, with respect to his own demands and wishes before he decides to migrate.

Therefore, it seems to be reasonable to introduce dynamic and quantitative measures $u(t)$, $v(t)$ for the attraction of the city c, and the hinterland h, respectively. These quantities $u(t)$, $v(t)$ are denoted as *dynamic utilities*.

This concept of *dynamic utility* to be used cannot be an ordinal one but must be cardinal since we have to assign numerical values to every utility. We have introduced the concept of 'dynamic utilities' in order to distinguish our approach from conventional procedures where the word utility is also used in a static context. In Section 3.1 for instance, we applied the *classical* concept of utility optimization to the phenomena of sudden urban growth. In contrast to the procedure of Section 3.1, our dynamic utilities are not maximized. Instead they constitute an essential part of the decision process. The relation of the dynamic utilities to the utilities used in static random choice theory will be discussed in Section 9.7.

There it will turn out that the dynamic master equation approach under certain restrictions ends up with the same structure of the stationary solution as the random utility theory. Therefore it seems to be justified to give the quantities $u(t)$, $v(t)$ the name "utilities" even if the system is not in an equilibrium state.

According to our general introduction of dynamics, we assume individual transition rates between city c and hinterland h of the following reasonable form:

$$p_{\uparrow}(n) = \nu \exp[u(n_c) - v(n_h)] \qquad \text{from } h \text{ to } c$$

$$p_{\downarrow}(n) = \nu \exp[u(n_h) - v(n_c)] \qquad \text{from } c \text{ to } h \tag{3.14}$$

where $u(n_c)$, $v(n_h)$ are the dynamic utility functions of the city and the hinterland, respectively. The parameter ν determines the time scale on which location changes occur and is therefore denoted as *mobility parameter.*

Using microeconomic concepts (Section 3.1) depending on the ingredients of the theory (income distribution, land use, transportation costs and production to mention a few) different utility functions $u(n)$, $v(n)$ could be obtained. However, all these utility functions end up or can be expanded into a power series with respect to the population sizes n_c or n_h , respectively. Of course, the coefficients and the signs of the coefficients of these power series sensitively depend on the underlying assumptions.

Therefore , we assume the following forms

$$u(n_c) = \sum_{i=0}^{I} \alpha_i n_c^i \qquad\qquad v(n_h) = \sum_{j=0}^{J} \beta_j n_h^j \tag{3.15}$$

which yield, making use of the relevant variable n introduced in (3.8), the *utility gain function*:

$$f(n) = u(n_c) - v(n_h) = \sum_{k=0}^{K} \gamma_k n^k \tag{3.16}$$

The coefficients γ_k are related to the coefficients α_i , β_j , N. For $f(n) > 0$, a utility gain is connected with the transition from h to c. This case leads to $p_{\uparrow}(n) > p_{\downarrow}(n)$.

3.2.4 The Stationary Solution of the Master Equation

Having chosen for the specific form of the transition probabilities it is now possible to make the model explicit by inserting (3.14-16) into all the equations.

We shall firstly determine the stationary solution $P_{st}(n)$ of the master equation (3.13). Stationarity means that all time derivatives vanish (see Chapter 9). It is obvious, that the stationary version of (3.13) is fulfilled, if the condition of *detailed balance* holds:

$$w_{\downarrow}(n+1)\, P_{st}(n+1) = w_{\uparrow}(n)\, P_{st}(n); \qquad -N \le n \le N \tag{3.17}$$

This condition means, that in the stationary case the probability flux from n to $(n + 1)$ is equal to the inverse flux from $(n + 1)$ to n. For our system (next-neighbour transitions only) condition (3.17) is always satisfied. Hence the exact stationary solution of the master equation (3.13) can be obtained by repeated application of (3.17) yielding:

$$P_{st}(n) = P_{st}(0) \prod_{m=1}^{n} \frac{w_{\uparrow}(m-1)}{w_{\downarrow}(m)}; \qquad +1 \le n \le N$$

$$P_{st}(n) = P_{st}(0) \prod_{m=-1}^{n} \frac{w_{\downarrow}(m+1)}{w_{\uparrow}(m)}; \qquad -N \le n \le -1\,. \tag{3.18}$$

The value $P_{st}(0)$ has to be determined by using the normalization condition (3.12). After inserting (3.14-16) into (3.18) the product expressions can be further evaluated and the explicit exact stationary solution to the master equation (3.13) is obtained:

$$P_{st}(n) = P_{st}(0) \frac{(N\,!)^2}{(2N)!} \binom{2N}{N+n} \exp[2F(n)] \tag{3.19}$$

where

$$F(n) = \sum_{m=1}^{n} f(m) = \sum_{m=1}^{n} \sum_{k=0}^{K} \gamma_k m^k . \tag{3.20}$$

A further evaluation of the binomial coefficients and the other factorials using *Stirling's* formula (Abramowitz and Stegun[6])

$$\ln(N!) \approx N \ln N - N \qquad \text{for } N \gg 1 \tag{3.21}$$

yields with the quasi-continuous variable

$$x = n/N ; \qquad -1 \leq x \leq 1 \tag{3.22}$$

a more convenient form for the stationary distribution function

$$P_{st}(Nx) = P_{st}(0) \exp[N\, U(x)] \tag{3.23}$$

with

$$U(x) = 2\, \Phi(x) - [(1+x)\ln(1+x) + (1-x)\ln(1-x)],$$

where we have used the abbreviation

$$\Phi(x) = \int_0^x f(x)dx = \int_0^x \sum_{k=0}^{K} \tilde{\gamma}_k x^k dx \tag{3.24}$$

and the scaled coefficients

$$\tilde{\gamma}_k = \gamma_k N^k . \tag{3.25}$$

The extrema x_m of this distribution function are determined by

$$\left.\frac{\partial U(x)}{\partial x}\right|_{x=x_m} = 2 f(x_m) - [\ln(1 + x_m) - \ln(1 - x_m)] = 0$$

or

$$x_m = \tanh f(x_m). \tag{3.26}$$

The extrema x_m is a maximum if

$$\left.\frac{\partial^2 U}{\partial x^2}\right|_{x_m} < 0 \qquad \text{or } f'(x_m) < \frac{1}{1 - x_m^2}$$

or a minimum if

$$\left.\frac{\partial^2 U}{\partial x^2}\right|_{x_m} > 0 \qquad \text{or } f'(x_m) > \frac{1}{1 - x_m^2}$$

Here we see once more the difference of our dynamic approach to the static utility concept. The stationary state of the system is in our theory of decision processes *not* characterized by equal utility functions

$$u(n_c) = v(n_h) \tag{3.27}$$

which would lead to

$$f(x) = 0. \tag{3.28}$$

but instead by (3.26).

Equal utilities mean that there is no utility gain connected with a change of the location for any individual. Then, of course, the individual transition rates $p_{\uparrow}(n) = p_{\downarrow}(n)$ are also equal. On the other hand, the configurational transition rates are *different*: $w_{\uparrow}(n) \neq w_{\downarrow}(n)$. The *stationary state* of this dynamic system is, however, characterized by *equal macroscopic migration fluxes* between city c and alternative sector h and vice versa or in other words by condition (3.17).

In order to illustrate the meaning of (3.26), we shall now discuss the case of special utility functions for the city c and the hinterland h:

$$u(n_c) = \alpha_0 + \alpha_1 n_c ; \qquad v(n_h) = \beta_0 + \beta_1 n_h . \tag{3.29}$$

For obvious reasons we may denote α_0 and β_0 as *preference parameter* of the city c and hinterland h, respectively, and α_1 and β_1 as *agglomeration parameter* of the city c and the hinterland h.

These utility functions lead to

$$f(x) = \tilde{\gamma}_0 + \tilde{\gamma}_1 x , \tag{3.30}$$

with

$$\tilde{\gamma}_0 = (\alpha_0 - \beta_0) + (\alpha_1 - \beta_1) N ; \qquad \tilde{\gamma}_1 = (\alpha_1 + \beta_1) N \tag{3.31}$$

and to the equation for the extrema

$$x_m = \tanh(\tilde{\gamma}_0 + \tilde{\gamma}_1 x_m). \tag{3.32}$$

which can now be discussed graphically by representing its right hand side and left hand side. Simultaneously, the corresponding probability distribution can be drawn schematically (see Figure 3.7).

In the case of $0 \le \tilde{\gamma}_1 \le 1$, that means for a moderate agglomeration parameter there exists always only one stationary solution of (3.32) corresponding to one maximum of the distribution function. This maximum is in the domain $-1 \le x_m < 0$ or $0 \le x_m \le 1$ for $\tilde{\gamma}_0 < 0$ or $\tilde{\gamma}_0 > 0$, respectively.

However, for a large agglomeration parameter $\tilde{\gamma}_1 > 1$, and if the absolute value of the preference parameter exceeds a critical value $|\tilde{\gamma}_0| > \tilde{\gamma}_{0c}$ there exists also only one stationary solution of (3.32). This case is depicted in Figures 3.7a and 3.7d.

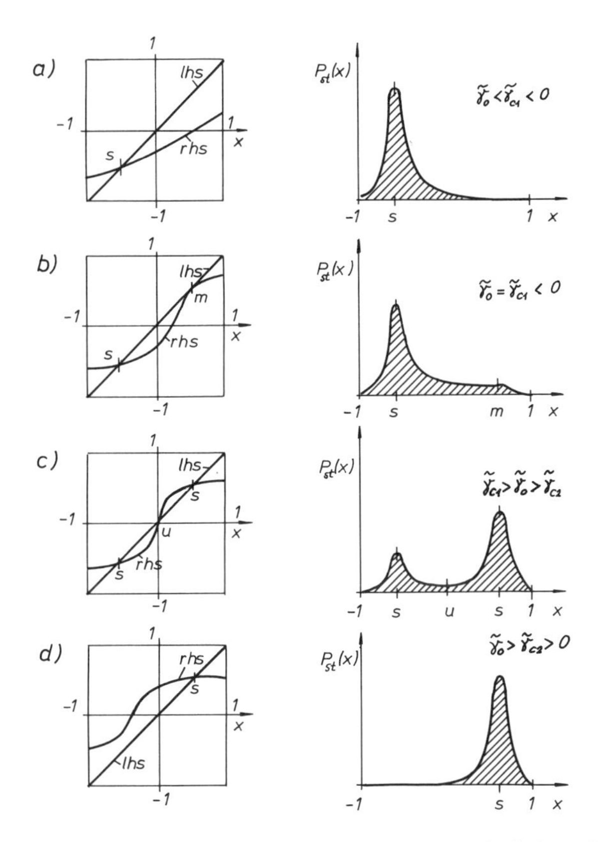

Figure 3.7: Stationary points and stationary probability distribution for different values of $\tilde{\gamma}_0$ and for $\tilde{\gamma}_1 > 1$

This critical value $\tilde{\gamma}_{0c}$ is determined by the equation

$$\cosh[\tilde{\gamma}_{0c} - \sqrt{\tilde{\gamma}_1(\tilde{\gamma}_1 - 1)}] = \tilde{\gamma}_1 \; : \qquad (3.33)$$

On the other hand, there exist three solutions of (3.32), corresponding to two maxima and one minimum, of the distribution, if $\tilde{\gamma}_1 > 1$, and if $|\tilde{\gamma}_0| < \tilde{\gamma}_{0c}$ according to Figure 3.7c.

The marginal case $\tilde{\gamma}_1 > 1$, $\tilde{\gamma}_0 = -\tilde{\gamma}_{0c}$ is shown in Figure 3.7b. We see that a second maximum of the distribution begins to evolve, if $\tilde{\gamma}_1 > 1$ and if $\tilde{\gamma}_0$ enters the domain $|\tilde{\gamma}_0| < \tilde{\gamma}_{0c}$.

Let us now interprete these cases: if there is no pronounced agglomeration trend, this means $0 < \tilde{\gamma}_1 < 1$, the maximum of the distribution function $P_{st}(n)$ (which describes the most probable partition of the population between city and hinterland) essentially depends on the relative preference parameter $\tilde{\gamma}_0 = (\alpha_0 - \beta_0) + (\alpha_1 - \beta_1)N$. According to the sign of this preference, the majority of the total population will live in the (rural) hinterland or in the city.

Starting from $\tilde{\gamma}_0 < 0$, a larger agglomeration trend $\tilde{\gamma}_1 > 1$ now changes the situation dramatically: at first a second maximum of the distribution appears (see Fig. 3.7b), and this will grow in its heigth relative to the old one, since $\tilde{\gamma}_0$ will also change its sign for larger $\tilde{\gamma}_1$. In effect, the most probable population partition has now shifted to an $x_m > 0$ with the majority living in the city (see Fig.3.7 c).

Although the foregoing results strictly apply to stationary situations with trend parameters $\tilde{\gamma}_0$, $\tilde{\gamma}_1$ kept constant, since we consider stationary distributions only, they can be generalized for slowly changing trend parameters: it can be shown that any initial distribution relaxes into the stationary one for given trend parameters. A change of trend parameters therefore means, that the distribution relaxes into the new stationary distribution belonging to the new trend parameters. Hence, if the evolution of trend parameters is slow enough, in comparison with this relaxation, the distribution will *adiabatically* follow by passing through a sequence of stationary states.

Applying this idea to the above discussed situations, we can reinterprete them in a dynamic manner: in many realistic situations (especially in Third

World countries) we start with a small agglomeration trend $\tilde{\gamma}_1 < 1$ and a preference $\tilde{\gamma}_0 < 0$ for the rural part of the country. Because the new economic development usually starts in the city area this may lead to a slow increase of the preference for the city area and of the agglomeration parameter $\tilde{\gamma}_1$ as well, so that at a certain time a situation $\tilde{\gamma}_1 > 1$ and $\tilde{\gamma}_0 > 0$ may be reached. This slow change of *control parameters* however, leads to a large and sudden shift of the population partition in favour of the city area: a migratory rush into the city area in the sense of a *phase transition* sets in (see Fig. 3.8).

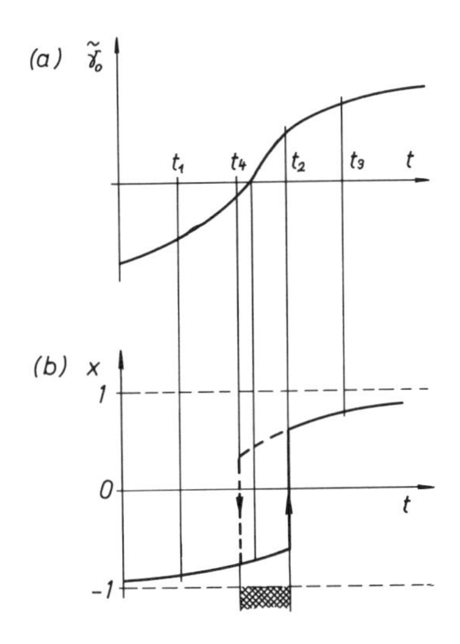

Figure 3.8: (a) expected change of $\tilde{\gamma}_0$, and (b) expected change of population share $x(t)$ (//// range of bistability)

As a main result of our theory we conclude that the expected change of the stationary points (and the shift of the probability distribution) necessary for explaining the phenomenon of *sudden urban growth* can be obtained with a *linear* utility function (3.30) within *this* model structure. However, the trend parameters have to fulfill the following requirements:

a) $\tilde{\gamma}_1 > 1$, since otherwise we do not obtain a first order phase transition as we expect in the case of sudden urban growth

b) $d\tilde{\gamma}_0 / dt > 0$, in the transition region. This means that the attraction of the city is growing at least in a certain interval of time

The main reasons of this sudden shift is the *self-accelerating* character of the agglomeration: the more people there are in the city, the more people are attracted by the city.

In the transition region we have to take into account the dynamics of the migration process. Therefore, we cannot expect that the probability distribution $P(n, t)$ agrees with $P_{st}(n)$. Hence, we have to consider the full dynamics of the master equation by numerical simulation or to derive appropriate equations of motion for the mean values and variances.

3.2.5 Equations of Motion for Mean Values and Variances

In most cases the full information contained in the probability distribution is not exploited in an empirical analysis. Often the mean values and their variances are already sufficient for the description of the dynamics of the system. Their definition and their equations of motion can be derived from the probability distribution and the master equation (compare Section 2.3.2).

The mean value of a function $g(n)$ is defined by:

$$\overline{g(n)} = \sum_{n=-N}^{N} g(n) P(n, t) . \tag{3.34}$$

The equations of motion for $\overline{n(t)}$ and $\overline{n(t)^2}$ follow from the master equation (3.13) in a straightforward manner by multiplying the right hand side and the left hand side with n and n^2, respectively and by summing over all states $-N \leq n \leq N$. A simple calculation subsequently leads to the following exact results:

$$\frac{d\,\overline{n(t)}}{d\,t} = \overline{w_{\uparrow}(n)} - \overline{w_{\downarrow}(n)} \tag{3.35}$$

and

$$\frac{d\,\overline{n(t)^2}}{d\,t} = 2\,\overline{n\,(w_{\uparrow}(n) - w_{\downarrow}(n))} + \overline{(w_{\uparrow}(n) + w_{\downarrow}(n))} \tag{3.36}$$

For essentially unimodal probability distributions one may approximate $\overline{g(n)}$ by $g(\bar{n})$ and thus arrive at self-contained although approximate equations of motion instead of (3.35) and (3.36). Introducing the scaled variable (3.22) $\bar{x} = \bar{n} / N$ and the variance

$$\sigma^2(t) = \overline{x^2(t)} - \overline{x(t)}^2; \qquad \Sigma^2(t) = N\,\sigma^2(t) \tag{3.37}$$

the following equations of motion can be derived from (3.35, 36):

$$\frac{d\,\overline{x(t)}}{dt} = K(\bar{x}), \tag{3.38}$$

$$\frac{d\,\overline{\Sigma^2(t)}}{dt} = 2\,K'(\bar{x})\,\overline{\Sigma^2(t)} + Q(\bar{x}) \tag{3.39}$$

with the abbreviations:

$$K(\bar{x}) = W_{\uparrow}(\bar{x}) - W_{\downarrow}(\bar{x}) \tag{3.40}$$

$$Q(\bar{x}) = W_{\uparrow}(\bar{x}) + W_{\downarrow}(\bar{x}) \tag{3.41}$$

where

$$W_{\uparrow}(\bar{x}) = N^{-1} w_{\uparrow}(\bar{n}) = \nu\,(1 - \bar{x})\exp[f(\bar{x})]$$

$$W_{\downarrow}(\bar{x}) = N^{-1} w_{\downarrow}(\bar{n}) = \nu\,(1 + \bar{x})\exp[-f(\bar{x})] \tag{3.42}$$

with

$$f(\bar{x}) = \sum_{k=0}^{K} \tilde{\gamma}_k \bar{x}^k. \tag{3.43}$$

Inserting (3.42) into (3.40) and (3.41) one obtains explicit forms for the *socio-economic driving force* $K(\bar{x})$ and the *fluctuation coefficient* $Q(\bar{x})$:

$$K(\bar{x}) = 2\nu\,[\sinh f(\bar{x}) - \bar{x}\cosh f(\bar{x})], \tag{3.44}$$

$$Q(\bar{x}) = 2\nu\,[\cosh f(\bar{x}) - \bar{x}\sinh f(\bar{x})]. \tag{3.45}$$

The mean value equation which describes the migration process between the city c and the hinterland h when $f(x)$ is the corresponding utility gain function then reads:

$$\frac{d\,\overline{x(t)}}{dt} = 2\nu\,[\sinh f(\bar{x}) - \bar{x}\cosh f(\bar{x})]. \tag{3.46}$$

This is a nonlinear differential equation for the *distribution variable* $\overline{x(t)}$. In the following we omit the bar for convenience, $\overline{x(t)} \to x(t)$.

The stationary solution of (3.46) can be read off immediately and agrees with the transcendental equation (3.26) for the most probable values x_m of the stationary probability distribution $P_{st}(n)$. This demonstrates the consistency of the mean value approach with the fully stochastic approach.

Under the assumption that the solution $\overline{x(t)}$ of (3.46) has been found, it is easy to find the explicit solution to (3.39)

$$\Sigma^2(t) = \Sigma^2[x(t)] = \Sigma_o\left[\frac{K(x(t))}{K(x(0))}\right]^2 + K^2(x(t))\int_{x(0)}^{x(t)} \frac{Q(x)}{K^3(x)}\,dx. \tag{3.47}$$

This solution is non-divergent only for $K(x(0)) \neq 0$ and describes the dynamics of the fluctuations in the evolution of $x(t)$, or in other words the stochastic nature of the underlying decision process.

If a city moves into a critical domain characterized by changes of the pattern of critical points and/or changes of the stability of critical points, in general an enhancement of fluctuations occurs anticipating the phase transition. Therefore, if we regard shocks in urban evolution as structural phase transitions of an urban system, the dynamics of the fluctuations can be used as an indicator of stability (see Fig. 3.9)

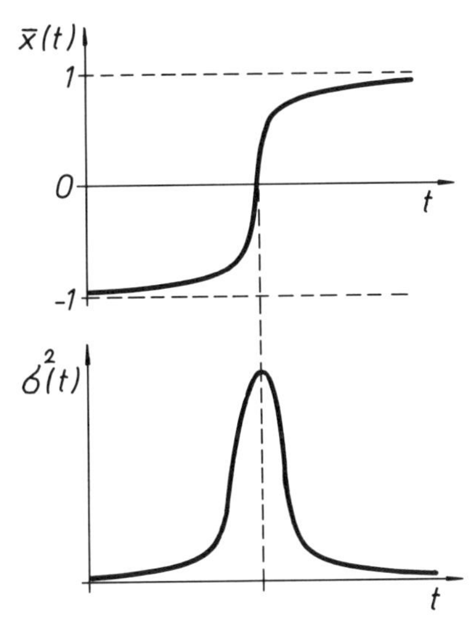

Figure 3.9: Expected evolution of the mean value $\bar{x}(t)$, and the variance $\sigma^2(t)$, in the case of sudden urban growth

3.2.6 Estimation of the Utility Function and Mobility

For a given set of trend parameters $\boldsymbol{x}(t) = (\nu(t), \gamma_k(t))$, the mean value equations (3.38, 39) can be used for calculating the time dependent solutions of $x(t)$, $\sigma^2(t)$ for a sequence of years. In general, however, the trend parameters are not known in advance. On the other hand, these parameters can be estimated for a sequence of years $t = 1, 2, \ldots, T$, if empirical data (denoted by the index e) are available.

We assume that the following empirical quantities can be observed (Table 3.1).

Table 3.1

Observed quantities per year (or census) for the migration process

time t	population numbers n_c	n_h	N^e	pop. share $x^e(t)$	number of transitions $W^e_\uparrow$	$W^e_\downarrow$
1	$n_c(1)$	$n_h(1)$	$N^e(1)$	$x^e(1)$	$W^e_\uparrow(1)$	$W^e_\downarrow(1)$
2	$n_c(2)$	$n_h(2)$	$N^e(2)$	$x^e(2)$	$W^e_\uparrow(2)$	$W^e_\downarrow(2)$
.	.	.	.	.	.	.
.	.	.	.	.	.	.
.	.	.	.	.	.	.
T	$n_c(T)$	$n_h(T)$	$N^e(T)$	$x^e(T)$	$W^e_\uparrow(T)$	$W^e_\downarrow(T)$

The theoretical expressions $W_\uparrow(x)$, $W_\downarrow(x)$ must be identified with the mean number of agents or individuals per unit of time $W^e_\uparrow(t)$, $W^e_\downarrow(t)$, migrating from the hinterland to the city, and vice versa. Therefore we have to match the theoretical expressions (3.42) to the empirical observations by an optimal choice of the trend parameters $\nu(t)$, $\gamma_k(t)$ (compare Section 2.3.1).

In the following paragraph we apply the log-linear estimation procedure. As we have seen, the dynamics of sudden urban growth can only be understood, if we assume that the population pattern undergoes a phase transition for which the stability of critical points changes with time. This can only happen if the trend parameters are also slowly time dependent quantities. Therefore, we must develop an estimation procedure allowing for the determination of slowly varying trend parameters.

In this sense, we minimize the sum of squared deviations between the theoretical expressions and the empirical data in a certain interval of time $(2\tau + 1)$:

$$F_t[\nu, \tilde{\boldsymbol{\gamma}}] = \sum_{t'=t-\tau}^{t+\tau} \{\ln w^e_{\uparrow}(t') - \ln[\nu(t')(1 - x^e(t'))\exp f(x^e(t')]\}^2$$
$$+ \sum_{t'=t-\tau}^{t+\tau} \{\ln w^e_{\downarrow}(t') - \ln[\nu(t')(1 + x^e(t'))\exp(- f(x^e(t'))]\}^2$$

with

$$f(x^e) = \sum_{k=0}^{K} \tilde{\gamma}_k(t) x^e(t)^k \tag{3.48}$$

Since the data base is often relatively bad, we improve the results by introducing a moving average in (3.48). The effects of uncertainties of the empirical observations can be considerably smoothed out then.

The necessary condition for $F_t[\nu, \tilde{\boldsymbol{\gamma}}]$ to assume its minimum is that its variation vanishes:

$$\delta F = \sum_{t'} \Big[\sum_{k} \frac{\partial F}{\partial \gamma_k} \delta\gamma_k(t) + \frac{\partial F}{\partial \nu} \delta\nu(t)\Big] = 0. \tag{3.49}$$

The variations $\delta\gamma_k$, $\delta\nu$ are considered to be independent of each other. Hence, each of the corresponding variational terms must vanish independently, yielding:

$$\frac{\partial F}{\partial \gamma_k(t)} = 0, \quad \text{for } k = 0, 1, 2, \ldots, K \quad \text{and } t = 1, 2, \ldots, T \tag{3.50}$$

and

$$\frac{\partial F}{\partial \nu(t)} = 0, \quad \text{for } t = 1, 2, \ldots, T. \tag{3.51}$$

After some calculations the evaluation of (3.50) leads to a set of linear equations for the time-dependent trend parameters $\tilde{\boldsymbol{\gamma}}(t)$ of the utility gain function:

$$\sum_{k=0}^{K} \tilde{\gamma}_k(t)\, C_{kl}(t) = D_l(t); \qquad l = 0, 1, 2, \ldots, K$$

$$t = 1, 2, 3, \ldots, T \tag{3.52}$$

where

$$C_{kl}(t) = C_{lk}(t) = \frac{1}{2\tau + 1} \sum_{t'=t-\tau}^{t+\tau} x^e(t')^{k+l} \tag{3.53}$$

and

$$D_l(t) = \frac{1}{2\tau + 1} \sum_{t'=t-\tau}^{t+\tau} \frac{x^e(t')}{2} \ln\left[\frac{(1 - x^e(t'))\, w^e_{\uparrow}(t')}{(1 + x^e(t'))\, w^e_{\downarrow}(t')}\right] \tag{3.54}$$

The mobility parameter of the population $\nu(t)$ is obtained by solving (3.51) and reads:

$$\nu(t) = \frac{1}{2\tau + 1} \sum_{t'=t-\tau}^{t+\tau} \sqrt{\frac{w^e_{\uparrow}(t') w^e_{\downarrow}(t')}{(1 - x^e(t')^2)}} \tag{3.55}$$

After determination of the trend parameters $\tilde{\boldsymbol{\gamma}}(t)$, $\nu(t)$, these can be inserted into the transition rates (3.42), which enter the equations of motion (3.38, 39). The latter then can be solved retrospectively or prospectively as discussed in the introduction.

3.2.7 Regression of Trend Parameters on Socio-Economic Data

The analysis of Section 3.2.6 yields a time series for the trend parameters $\boldsymbol{\varkappa}(t) = \{\nu(t), \tilde{\boldsymbol{\gamma}}(t)\}$, $t = 1, 2, \ldots, T$. A crucial question not answered hitherto is the determination of the relevant socio-economic driving factors - called key-factors - of sudden urban growth.

We shall assume that there exists a set of properly standardized socio-economic factors $\boldsymbol{\Omega}(t) = \{\Omega_\alpha(t)\}$, $\alpha = 1, 2, \ldots, A$ which seem to be related to the migratory dynamics (e.g., income, vacancies available, traffic, infrastucture, schools and rents). A tentative linear relation between the trend parameters and the socio-economic factors can be assumed:

$$\boldsymbol{x}(t) = \bar{a}\, \boldsymbol{\Omega}(t), \tag{3.56}$$

where $\bar{a} = \{\bar{a}_{i\alpha}\}$ is the matrix of the corresponding influence factors. The optimal choice of the influence matrix $\bar{a}$ can be found by regression methods (see Section 2.3.3) and will not be further outlined here.

Chapter 4

Intra - Urban Migration

The following model of intra-urban population dynamics (for a detailed description see Haag and Dendrinos[1] and Dendrinos and Haag[2]) is partially similar to the preceding one of shocks in urban evolution. On the other hand it differs in one important point: the trend parameters governing the population dynamics are directly linked to economic quantities, namely rents. These rents are set by landowners. Therefore they can be treated as dynamic and stochastic quantities on a similar basis as the population numbers; this means they can be directly integrated into the master equation and mean value equation framework. Hence, the rents here are endogeneous, fully dynamic variables coupled to the dynamics of population numbers. Consequently we shall end up with a master equation for a probability distribution over population numbers and rents per unit area of land.

4.1 Introduction

It it our purpose to provide an economically motivated theory which describes selectively certain problems in intra-urban location. Phenomena of multiple dynamical states and bifurcation behaviour are captured, together with some aspects of instabilities characterizing intra-urban allocation.

Continuous and sudden changes in the internal structure of cities have been widely recorded as well as empirically confirmed (compare the exhaustive literature reported in Dendrinos[3]). The incidence of suburbanization in North

America, Western Europe, and other regions has attracted considerable interest. In particular slum formation and the tipping of urban neighbourhoods in relatively short time periods from one type of residence to another are events widely documented in the United States.

Sudden transitions have occurred in communities of central cities from white, middle-income households to predominently black, low-to middle-income families. A new phenomenon, referred to as *gentrification* has also been recently identified in certain US metropolitan areas. Gentrification is understood as a transition whereby white families repossess dwellings previously owned by black households or in other words the transition from white to black residences in a particular neighbourhood is being reversed. It is the first indication of a cyclical movement in the residential sector by type of households.

There exists a thorough literature on intra-urban theory and models. We mention some major work here only. Von Thünen's [4] important initiating work by has been considerably extended by Beckmann, [5] Alonso,[6] Mills,[7] Miyao[8] and others. These works are based on intra-urban economic theory. A partial survey on land use and transportation models can be found in Pack.[9] The dynamic spatial interaction models of Wilson,[10,11] and Allen et al[12,13] can be characterized as *operational*. Forrester's[14] urban dynamics model, however, belongs to the class of large scale models and is difficult to analyse in parameter space, since exogeneous shocks are randomly imposed in the system of differential equations.

A different approach is attempted here in modelling the basic features of intra-urban ecology. A stochastic master equation formulation of the central city - suburb, rent - density dynamics will be provided. We will test our intra-urban dynamic theory on empirical data of twelve U.S. Standard Metropolitan Statistical Areas (SMSAs). The model is also used for predicting the intra-urban rent and density distribution of these SMSAs for 1990.

4.2 A Stochastic Model on Intra - Urban Dynamics

We assume a particular intra-urban configuration by subdividing the urban setting into two zones, namely a central city c and a surrounding suburban ring s (see Fig. 4.1).

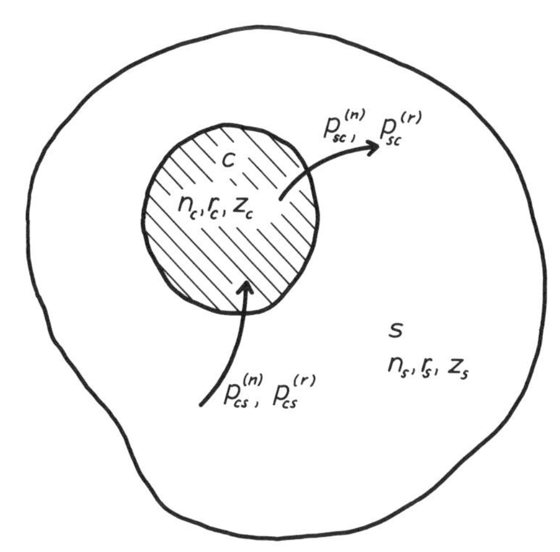

Figure 4.1: The intra-urban configuration: the central city c, and the suburban ring s

However, the model is general in so far as *any* particular spatial breakdown into two zones can be treated with the same formalism. The total area of land available in c and s is z_c and z_s, respectively, and the respective population densities are n_c and n_s. There are two sectors in this economy, buyers of land with *bid rents* $\tilde{r}_c$ and $\tilde{r}_s$, and sellers of land with *asking rents* r_c and r_s per unit area of land in the two zones correspondingly. It is convenient to introduce the following variables:

$$n_c z_c + n_s z_s = 2N; \qquad n_c z_c - n_s z_s = 2n; \tag{4.1}$$

yielding

$$n_c z_c = N + n; \qquad n_s z_s = N - n; \tag{4.2}$$

and

$$r_c z_c + r_s z_s = 2R; \qquad r_c z_c - r_s z_s = 2r; \tag{4.3}$$

yielding

$$r_c z_c = R + r; \qquad r_s z_s = R - r \tag{4.4}$$

and

$$- N \leq n \leq N \qquad - R \leq r \leq R. \tag{4.5}$$

As a consequence of the definitions of n_c, n_s, r_c, r_s and z_c, z_s, N is half of the total urban population and R is half of the total land rent payment received by the absentee land owners.

In order not to complicate matters we make the assumption that N, R and the total urban land $Z = z_c + z_s$ remain constants in time. The quantities n and r designate the deviation from a uniform allocation of population and rental value to the two zones. We employ a normalization of rental and population densities later, thus view the *relative dynamic allocation* of population and rental payments.

4.2.1 The Configurational Transition Rates

At the *microlevel* the urban system is heterogeneous. It consists in reality of a number of distinct agents, renters, and suppliers of land improvements. These agents differ in preferences, incomes, profit functions, and any other relevant attribute. We can, however, make the following (testable) hypothesis for the purposes of studying the urban system, that the $2N$ agents operate *as if* all agents in the demand sector act according to a probability function depicting the decision behaviour of a *representative agent*. An equivalent "*as if* " hypothesis can also be made for the supply sector.

The demand side

Firstly, population dynamics can be treated along the lines already discussed in Chapter 3. The resident in one of the zones has a certain willingness to move to the other zone. This willingness is a function of the estimated difference of *utilities* of both zones. It is expressed by individual transition rates of the form:

$$p_{cs}^{(n)} = \alpha \exp(u_c - u_s) \qquad \text{from } s \text{ to } c$$

$$p_{sc}^{(n)} = \alpha \exp(u_s - u_c) \qquad \text{from } c \text{ to } s. \tag{4.6}$$

The utilities u_c , u_s in turn are functions of the respective densities and rents, and α is a time scaling (mobility) parameter.

A plausible assumption for the utilities is:

$$u_c = a_1 (\tilde{r}_c - r_c) + a_2 n_c$$

$$u_s = a_1 (\tilde{r}_s - r_s) + a_2 n_s \tag{4.7}$$

The last term of the right hand side of (4.7) describes agglomeration or deglomeration effects: a_2 is either positive or negative depending on assumptions made regarding aversion or attraction to high density living. The first term of the right hand side describes (with positive a_1) the increasing utility of a zone, if the bid rent exceeds the asked rent. We treat the asked rent r_c , r_s as a dynamic variable, whereas we keep constant the bid rents $\tilde{r}_c$, $\tilde{r}_s$ for the period of time considered. Therefore, rental comparative advantages are one element in this model that drive up the probability of making a move.

The individual transition rates (4.6) lead to configurational transition rates for the transition $n \rightarrow (n + 1)$ and $n \rightarrow (n - 1)$, where n is half the difference of population numbers in both zones (4.1):

$$w[n+1, r; n, r] = n_s z_s p_{cs}^{(n)} = \alpha n_s z_s \exp(u_c - u_s)$$

$$w[n-1, r; n, r] = n_c z_c p_{sc}^{(n)} = \alpha n_c z_c \exp(u_s - u_c) \qquad (4.8)$$

In (4.8) it is assumed, that the population movement n is not simultaneously connected with a transition in rental distribution r.

The supply side

Secondly, we have to consider the dynamics of the rent variable r (see 4.3). Since increases or decreases of asking rents in both zones are caused in single steps by individual land suppliers (that means, land suppliers are the agents and decide about the rent level), it is justified to treat the relative rental payment variable r as a stochastic variable, too. The units can be adjusted in such a way that the typical single change in the rent level lead to transitions $r \rightarrow (r+1)$ or $r \rightarrow (r-1)$, respectively. For these steps there exist individual transition probabilities of single suppliers:

$$p_{cs}^{(r)} = \beta \exp(v_{cs}) \qquad \text{transfer from } s \text{ to } c$$

$$p_{sc}^{(r)} = \beta \exp(v_{sc}) \qquad \text{transfer from } c \text{ to } s. \qquad (4.9)$$

We assume, that the *transfer utilities* v_{cs}, v_{sc} for the land suppliers depend on the *net excess demand*, defined as $[(n_c - \tilde{n}_c) - (n_s - \tilde{n}_s)]$. Presumably, suppliers are motivated to increase the asking rent at each time period t when net excess demand is observed. If $(n_i - \tilde{n}_i)$ $i = c, s$ is the difference between current density level n_i and occupancy level supplied $\tilde{n}_i$, it is reasonable to assume:

$$v_{cs} = a_3[(n_c - \tilde{n}_c) - (n_s - \tilde{n}_s)]$$

$$v_{cs} = - v_{sc} \qquad (4.10)$$

Since the total number of independent agents in zone c or s can be roughly assumed as proportional to the total rents $r_c z_c$ or $r_s z_s$, we obtain total transition rates for the rent adjustment process $r \to (r + 1)$ and $r \to (r - 1)$ of the following form:

$$w[n, r+1; n, r] = r_s z_s p_{cs}^{(r)} = \beta\, r_s z_s \exp(v_{cs})$$

$$w[n, r-1; n, r] = r_c z_c p_{sc}^{(r)} = \beta\, r_c z_c \exp(v_{sc}) \qquad (4.11)$$

where β is a measure of the intensity of the response of the individual land owner to an imbalance in net excess demand.

Introducing more convenient coefficients, the total configurational transition rates (4.8) and (4.11) can finally be written in form:

$$w[n+1, r; n, r] = \alpha\, (N - n)\exp(b_0 - b_1 r + b_2 n)$$

$$w[n-1, r; n, r] = \alpha\, (N + n)\exp[-(b_0 - b_1 r + b_2 n)]$$

$$w[n, r+1; n, r] = \beta\, (R - r)\exp(-b_3 + b_4 n)$$

$$w[n, r-1; n, r] = \beta\, (R + r)\exp[-(-b_3 + b_4 n)] \qquad (4.12)$$

where

$$b_0 = a_1(\tilde{r}_c - \tilde{r}_s) - (a_1 R - a_2 N)(1/z_c - 1/z_s) \qquad (4.13)$$

$$b_1 = a_1(1/z_c + 1/z_s) > 0 \qquad (4.14)$$

$$b_2 = a_2(1/z_c + 1/z_s) \tag{4.15}$$

$$b_3 = a_3[(\tilde{n}_c - \tilde{n}_s) + N(1/z_s - 1/z_c)] \tag{4.16}$$

$$b_4 = a_3(1/z_c + 1/z_s) > 0 \tag{4.17}$$

The parameters a_1 and a_3 (both always positive) designate the propensity to respond to comparative advantages in rental value at the demand and supply side respectively. According to the way the metropolitan area is partitioned (splitting of Z) these propensities must be adjusted . The parameters b_2 could be either positive or negative if attraction or aversion to high density living prevails.

4.2.2 The Master Equation

The master equation for the probability distribution $P(n, r, t)$ for population, n, and rental value, r, at any time period t, can now be set up in standard manner (see Chapters 2, 9) as a probability rate equation:

$$\begin{aligned} \frac{dP(n, r, t)}{dt} &= w[n,r;n+1,r]P(n+1,r,t) - w[n-1,r;n,r]P(n,r,t) \\ &+ w[n,r;n-1,r]P(n-1,r,t) - w[n+1,r;n,r]P(n,r,t) \\ &+ w[n,r;n,r+1]P(n,r+1,t) - w[n,r-1;n,r]P(n,r.t) \\ &+ w[n,r;n,r-1]P(n,r-1,t) - w[n,r+1;n,r]P(n,r,t) \end{aligned} \tag{4.18}$$

where the probability distribution has to be normalized accordingly

$$\sum_{r=-R}^{r=R} \sum_{n=-N}^{n=N} P(n,r,t) = 1 . \tag{4.19}$$

4.2.3 The Stationary Solution of the Master Equation

The stationary solution of the master equation (4.18), $P_{st}(n,r)$ provides the most probable population and rental values, as well as the fluctuations around it.

Before investigating the form of the stationary state, the detailed balance condition is examined (see Appendix Chapter 9.4, 9.5). It is found, that in the (n,r) space the condition does not hold, for $b_1 \neq - b_4$. But $b_1 \neq b_4$ is always the case, because both coefficients must be positive in order to make economic sense. Coefficients b_1 and b_4 depict positive propensities to move and to increase the asked rent when population faces rental advantages and suppliers face excess demand, correspondingly.

Although the condition of detailed balance does not hold for the present model, an approximate stationary solution can be constructed (see Appendix, Chapter 9.5). It reads:

$$P_{st}(n,r) = C \binom{2N}{N+n}\binom{2R}{R+r} \exp g(n,r) \tag{4.20}$$

with

$$g(n,r) = 2b_0 n - 2b_3 r + b_2 n^2 + (b_4 - b_1) n\, r . \tag{4.21}$$

The extrema $(\hat{n}, \hat{r})$ of the stationary distribution $P_{st}(n,r)$ are now determined by the conditions:

$$\left.\frac{\partial P_{st}(n,r)}{\partial n}\right|_{\hat{n},\hat{r}} = 0 ; \qquad \left.\frac{\partial P_{st}(n,r)}{\partial r}\right|_{\hat{n},\hat{r}} = 0 ; \tag{4.22}$$

which imply:

$$\frac{\hat{n}}{N} = \tanh[b_0 + b_2 \hat{n} + \hat{r}\,(b_4 - b_1)/2] \tag{4.23}$$

$$\frac{\hat{r}}{R} = \tanh[-b_3 + \hat{n}\,(b_4 - b_1)/2]. \tag{4.24}$$

Conditions (4.23, 24) provide two transcendental equations in population and rental value distribution that precisely determine the maxima and minima of the stationary distribution.

4.2.4 The Mean Value Equations

We shall now introduce the mean values and derive their equations of motion. The mean value $\overline{f(n,r)}$ of a function $f(n,r)$ of the variables n and r is generally defined as:

$$\overline{f(n,r)} = \sum_{r=-R}^{r=R} \sum_{n=-N}^{n=N} f(n,r)P(n,r,t) . \tag{4.25}$$

The mean values $\overline{n(t)}$ and $\overline{r(t)}$ are special cases of (4.25). Making use of the master equation (4.18), equations of motion for the mean values can now be derived in the same way as indicated in Chapter 2.

These mean value equations finally read:

$$\frac{d\overline{n(t)}}{dt} = \overline{k_n(n, r)} \tag{4.26}$$

$$\frac{d\overline{r(t)}}{dt} = \overline{k_r(n, r)}. \tag{4.27}$$

with

$$k_n(n, r) = w[n+1,r;n,r] - w[n-1,r;n,r]$$

$$k_r(n, r) = w[n,r+1;n,r] - w[n,r-1;n,r] \tag{4.28}$$

Equations of motion for the variances can also be derived, but will be skipped here (for details see Haag and Dendrinos[1]).

The *local forces* $k_n(n, r)$, $k_r(n, r)$ describe the net effect of the pull and push factors in population movement from any particular distribution and in rental value transfer from any particular rental allocation scheme.

The exact equations (4.26, 27) become selfcontained, though approximate equations if the assumption:

$$\overline{k_n(n, r)} \approx k_n(\bar{n}, \bar{r}) \qquad \overline{k_r(n, r)} \approx k_r(\bar{n}, \bar{r}) \tag{4.29}$$

holds. Normalizing $\bar{n}$ and $\bar{r}$ by

$$x = \bar{n}/N\,; \qquad -1 \leq x \leq 1$$

$$y = \bar{r}/R\,; \qquad -1 \leq y \leq 1 \tag{4.30}$$

and introducing the scaled coefficients $\tilde{b}_i$ and scaled time τ

$$\tilde{b}_1 = b_1 R\,; \quad \tilde{b}_2 = b_2 N\,; \quad \tilde{b}_4 = b_4 N\,; \quad \tau = 2\alpha t\,; \quad \gamma = \beta/\alpha \tag{4.31}$$

we obtain from (4.26, 27) with (4.12), (4.28) the explicit equations of motion for mean values (deterministic equations) on relative dynamic zonal density and rent distributions:

$$\frac{dx}{d\tau} = \sinh(b_0 - \tilde{b}_1 y + \tilde{b}_2 x) - x\cosh(b_0 - \tilde{b}_1 x + \tilde{b}_2 x) \tag{4.32}$$

$$\frac{dy}{d\tau} = \gamma\,[\sinh(-b_3 + \tilde{b}_4 x) - y\cosh(-b_3 + \tilde{b}_4 x)] \tag{4.33}$$

The global structure of these nonlinear deterministic equations of our intra-urban allocation model depends on the number and structure of their singular (stationary) points. The singular points x^*, y^* are defined by the simultaneous vanishing of the time derivatives, left hand side, of (4.32, 33). This yields the equations:

$$x^{*} = \tanh(b_0 - \tilde{b}_1 y^{*} + \tilde{b}_2 x^{*}) \tag{4.34}$$

$$y^{*} = \tanh(-b_3 + \tilde{b}_4 x^{*}) \tag{4.35}$$

which are independent of γ and result in the transcendental equation:

$$x^{*} = G(x^{*}), \tag{4.36}$$

with

$$G(x^{*}) = \tanh[b_0 + \tilde{b}_2 x^{*} - \tilde{b}_1 \tanh(-b_3 + \tilde{b}_4 x^{*})]. \tag{4.37}$$

To examine the qualitative properties of the dynamical system we consider the special case $b_0 = b_3 = 0$. If the right hand side of (4.37) has a slope at point $x^{*} = 0$ less than one, then there is only one critical value, satisfying (4.37), the origin. If, on the other hand, the slope is larger than one, there exist at least three critical points, with the origin being one of them. Thus, in this special case, it is possible that a transition from one into another equilibrium point could occur. The condition distinguishing the cases is

$$\left.\frac{d\,G(x)}{dx}\right|_{x^{*}=0} = \tilde{b}_2 - \tilde{b}_1 \tilde{b}_4 = \begin{cases} < 1 & \text{one critical point} \\ \phantom{<}\, 1 & \text{bifurcation} \\ > 1 & \text{multiple of critical points} \end{cases} \tag{4.38}$$

In the case where b_0 and/or b_3 are not zero the equilibrium points will still lie in the neighbourhood of the ones depicted in the special case. Although the magnitude of b_0 and b_3 affects the location of the singular points, it does not affect the *nature* of the dynamic equilibria.

In order to examine the nature of the critical points we linearize the system of simultaneous differential equations (4.32, 33) near the critical (stationary) points. We will not discuss this here, but remark, that there exists a stable limit cycle in a narrow range when:

$$1 + \gamma < \tilde{b}_2 < 1 + \tilde{b}_1 \tilde{b}_4 \tag{4.39}$$

is fulfilled (see /4.1, 4.2/). But if β is greater than α, then $\gamma > \tilde{b}_1 \tilde{b}_4$ holds, and no stable limit cycle is feasible. If $\gamma < \tilde{b}_1 \tilde{b}_4$ but $\tilde{b}_2 > 1 + \tilde{b}_1 \tilde{b}_4$, then there exists an unstable limit cycle and two stable sinks, symmetric to the origin $x^* = y^* = 0$.

4.2.5 The Parameter Estimation Procedure

The core of the theoretical model proposed, is the deterministic equations (4.32, 33) containing parameters (α, β, b_0 , $\tilde{b}_1$, $\tilde{b}_2$, b_3 and $\tilde{b}_4$). However, since the theoretical transition probabilities (4.12), which reappear in the mean value equations (4.32, 4.33) and which can be expressed in the normalized variables x, y by:

$$\mu_{cs}(x,y) = N^{-1} w[n+1,r;n,r] = \alpha\,(1 - x)\exp(b_0 - \tilde{b}_1 y + \tilde{b}_2 x)$$

$$\mu_{sc}(x,y) = N^{-1} w[n-1,r;n,r] = \alpha\,(1 + x)\exp[-(b_0 - \tilde{b}_1 y + \tilde{b}_2 x)]$$

$$\nu_{cs}(x,y) = N^{-1} w[n,r+1;n,r] = \beta\,(1 - x)\exp(-b_3 + \tilde{b}_4 x)$$

$$\nu_{sc}(x,y) = N^{-1} w[n,r-1;n,r] = \beta\,(1 + x)\exp[-(-b_3 + \tilde{b}_4 x)] \tag{4.40}$$

can be compared with the corresponding empiric quantities μ^e_{sc}, μ^e_{cs} if such data are available f or selected years $t = 1, 2, \ldots, T$, the estimation of the parameters of the model is manageable.

Under the assumption that the trend parameters do not vary over time and space (remaining identical for all SMSAs), there are potentially enough time series data available to estimate statistically the values of α, b_0 , $\tilde{b}_1$, and $\tilde{b}_2$. Owing to lack of data on ν^e_{cs} and ν^e_{sc} , the change of rental value at any particular time, the parameters β, b_3 and $\tilde{b}_4$ cannot be estimated directly. But through comparison with the numerical simulation of the equations we can indirectly also estimate β, b_3, and $\tilde{b}_4$.

In the following Section a short description of the statistical estimation process involved in computing the trend parameters α, b_0, $\tilde{b}_1$, $\tilde{b}_2$ is added. We proceed along the lines introduced in Chapter 2. A log-linear estimation of the trend parameters with:

$$F(\alpha,b_0,\tilde{b}_1,\tilde{b}_2) = \sum_{t=1}^{T}\{[\ln\mu^e_{cs} - \ln\mu_{cs}(x,y)]^2 + [\ln\mu^e_{sc} - \ln\mu_{sc}(x,y)]^2\} = \min ! \tag{4.41}$$

finally yields inserting (4.40) into (4.41), the mobility of the population:

$$\alpha = \frac{1}{T}\sum_{t=1}^{T}\sqrt{\frac{\mu^e_{cs}(t)\,\mu^e_{sc}(t)}{1 - x^e(t)^2}} \tag{4.42}$$

and the other trend parameters

$$\tilde{b}_1 = \frac{-\sigma_{xx}\sigma_{cy} + \sigma_{xy}\sigma_{cx}}{(\sigma_{xx}\sigma_{yy} - \sigma_{xy}\sigma_{xy})} \qquad \tilde{b}_2 = \frac{\sigma_{yy}\sigma_{cx} - \sigma_{xy}\sigma_{cy}}{(\sigma_{xx}\sigma_{yy} - \sigma_{xy}\sigma_{xy})}$$

$$b_0 = \bar{c} + \tilde{b}_1\bar{y} - \tilde{b}_2\bar{x} \tag{4.43}$$

where the variances have been introduced:

$$\sigma_{xx} = (\overline{x^2} - \bar{x}^2); \qquad \sigma_{xy} = (\overline{xy} - \bar{x}\,\bar{y});$$

$$\sigma_{yy} = (\overline{y^2} - \bar{y}^2); \qquad \sigma_{cy} = (\overline{cy} - \bar{c}\,\bar{y});$$

$$\sigma_{cx} = (\overline{cx} - \bar{c}\,\bar{x}); \tag{4.44}$$

and where $c(t)$ is the abbreviation:

$$c(t) = \frac{1}{2} \ln \frac{(1+x^e)\mu_{cs}^e}{(1-x^e)\mu_{sc}^e}. \tag{4.45}$$

The *bar* in (4.43, 44) indicates that we have to take the temporal mean value of the corresponding empirically known quantities (We have omitted the index e in (4.44), in order to simplify the notation).

As mentioned before, the remaining parameters β, b_3 , $\tilde{b}_4$ can be found by matching the computer solution of the mean value equations (4.32, 33) to the empirical data set.

4.2.6 Empirical Testing of the Land Use Density-Rent Model

The results presented here draw from Dendrinos and Haag's work. The time-series data from twelve US SMSAs (see Table 4.2) were used for the 1950 to 1980 period. Aggregate data on population density and average rent per unit area of land were employed, at the central cities and suburb levels of metropolitan areas. We also used data on intra-urban population migration counts.

Table 4.1

The optimal values of the trend parameters b_0, $\tilde{b}_1$, $\tilde{b}_2$, b_3 and $\tilde{b}_4$ and their variances

trend parameter	estimated value	variance
b_0	0.20	0.02
$\tilde{b}_1$	0.68	$8.5 \cdot 10^{-4}$
$\tilde{b}_2$	0.83	$9.1 \cdot 10^{-3}$
b_3	- 1.00	$4 \cdot 10^{-2}$
$\tilde{b}_4$	1.50	$9 \cdot 10^{-2}$

For reasons of data limitations, it was necessary to assume that a number of models' parameters do not vary from one metropolitan area to another and remain constant over time. Using this assumption, the optimal values of the five trend parameters b_0, $\tilde{b}_1$, $\tilde{b}_2$, b_3, and $\tilde{b}_4$ for the period of time 1950 to

The remaining trend parameters α, β and the predicted values of x and y in 1990 are listed in Table 4.2.

Table 4.2

Estimated trend parameters for twelve US SMSAs. The 1990 values are indicated, together with the stationary state

SMSAs	α	β	$x(1990)$	$y(1990)$
Buffalo	0.021	0.0126	- 0.52	0.84
Philadelphia	0.017	0.0057	- 0.37	0.79
Pittsburgh	0.018	0.0014	- 0.67	0.85
Portland	0.024	0.0020	- 0.63	0.87
San Diego	0.034	0.0019	- 0.71	0.89
Spokane	0.017	0.0037	- 0.24	0.92
Atlantic City	0.031	0.0312	- 0.63	0.73
Erie	0.012	0.0122	- 0.24	0.92
Miami	0.031	0.0047	- 0.64	0.75
Altona	0.010	0.0015	- 0.24	0.93
Omaha	0.030	0.0090	- 0.48	0.74
Trenton	0.031	0.0155	- 0.46	0.56

stationary values $x^* = -0.369$; $y^* = 0.419$

In Figures 4.2 and 4.3 the simulated dynamic paths of ten (of the twelve) SMSAs are shown. The dynamic paths of Buffalo and Eri are not plotted because of their proximity to those of Miami and Spokane. In both Figures

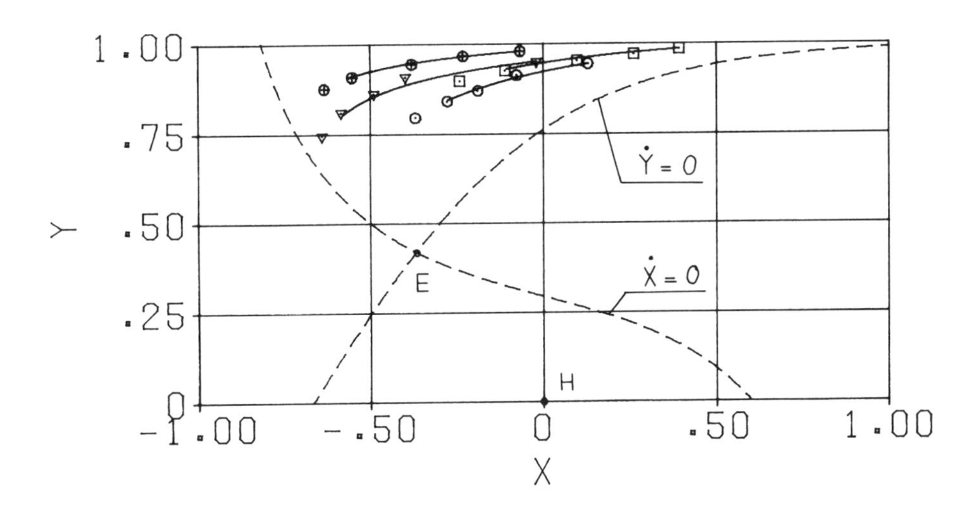

Figure 4.2: Simulated trajectories and actual counts for four SMSAs. ⊕ San Diego, ⊙ Trenton, ▽ Pittsburgh, △ Altona, ◇ Atlantic City, ⊡ Omaha.

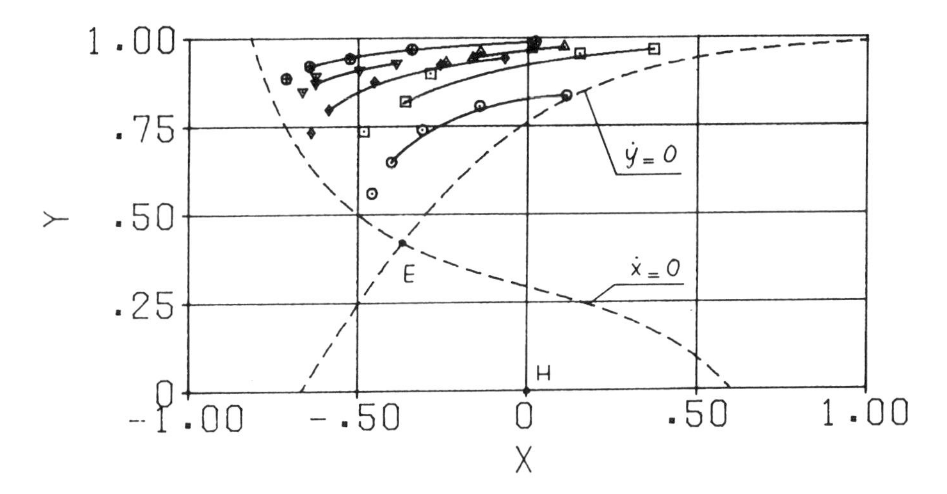

Figure 4.3 : Simulated trajectories and actual counts for four SMSAs. ⊕ Portland, ⊡ Spokane, ⊙ Philadelphia, ▽ Miami

the two isoclines ($\dot{x} = 0, \dot{y} = 0$) are depicted also (by dashes), together with the steady state E, which is a stable stationary point of the equations (4.32, 33) for parameters (4.46) and which will be reached for $t \to \infty$ under model conditions remaining constant. Point H represents the even population and rental value distribution between central city and suburbs of all SMSAs. The observed motion towards the left along the x-axis implies relative increase of rental values in the suburbs. The plotted symbols of the SMSAs belong to 1950, 1960, 1970 and 1980 counts (numbered from the right to the left). The isolated points to the left show the 1990 projection.

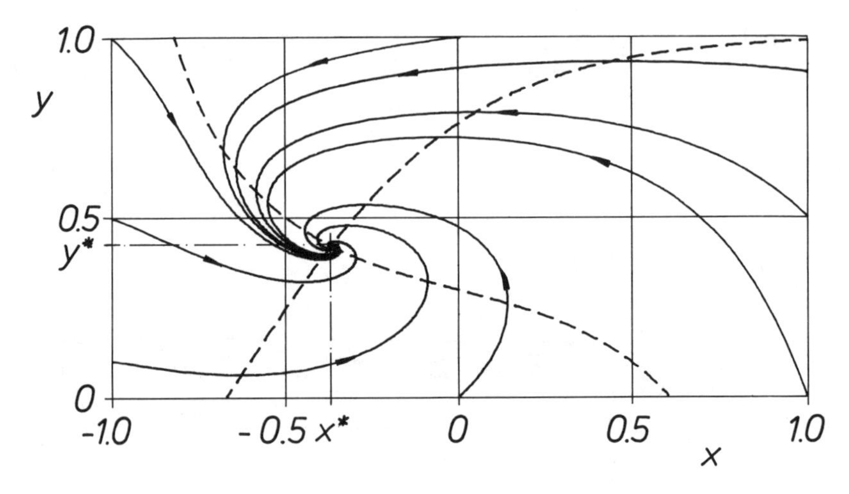

Figure 4.4: The spiraling sink nature of the unique intra-urban equilibrium. Parameter values generating this phase portrait are $b_0 = 0.2$, $\tilde{b}_1 = 0.68$, $\tilde{b}_2 = 0.82$, $b_3 = -1.0$, $\tilde{b}_4 = 1.5$, $\alpha = 0.031$, $\beta = 0.0047$

From Tables 4.1, 4.2 and Figures 4.2 to 4.4 some conclusions can be drawn. Primary among these is that the deterministic mean value equations of population rent interaction proposed fit the empirical data quite well. The simulation runs further indicate that although during the calibration period (1950-80) no oscillations are present by 1990 the population shift to the suburbs is expected to be reversed. Suburbs will continue to enlarge their share of land value. The relative flexibility of changing rental value seems to be lower than that of relative population mobility, because we obtain in general for all SMSAs $\alpha > \beta$. The size of a metropolitan area does affect their place in the phase portrait: the larger the size, the closer towards the steady state is the dynamic path. This can also be detected in the value of relative population mobility α: the larger the SMSA in population, the lower the value of α. The steady state is characterized by the majority of the SMSAs population living in the suburbs.

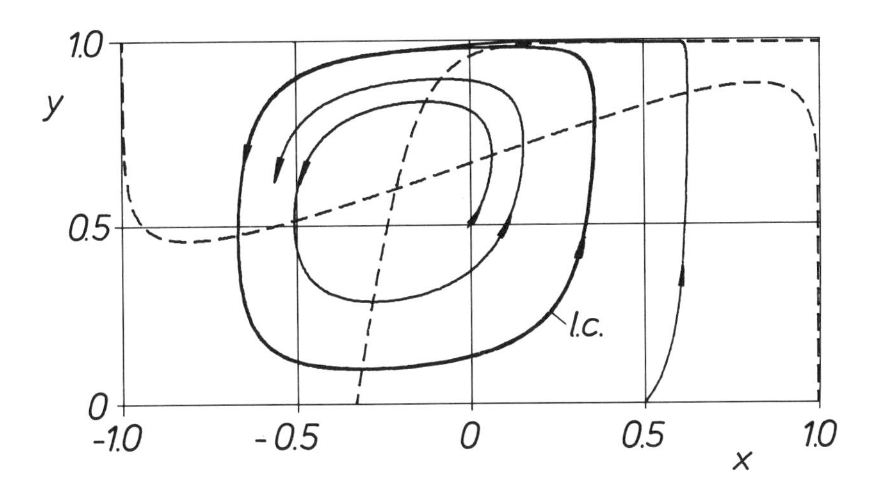

Figure 4.5: The possibility for a stable limit cycle (l.c.). Parameter values responsible for this phase portrait are $b_0 = 4.0$, $\tilde{b}_1 = 6.0$, $\tilde{b}_2 = 3.0$, $b_3 = -2.0$, $\tilde{b}_4 = 6.0$, $\alpha = 0.01$, $\beta = 0.01$

All conclusions of course critically hinge upon the premise that the parameters shared by all SMSAs (one may say the *environment*) will remain constant over time. This is rather unrealistic for very long time horizons. However, the model is capable of predicting that the qualitative nature of the dynamic equilibria will always be of sink spiral type. There is a very small chance that certain thresholds may be reached where a phase transition whould occur so that the stable focus will be transformed into a stable limit cycle (see also Figure 4.5). The relative remoteness of the current trend parameter values from such a set is an indication of the unlikeness of stable oscillatory behaviour in population density - rent value interaction.

Chapter 5

Inter - Regional Migration

In this Chapter we proceed to the more general case, namely the migration of individuals between different regions of a country. We start by considering the microlevel of individual decision processes to migrate. On the macrolevel rich quantitative empiric material about migration is available in many countries for a considerably long number of years. These data provide a rather good empiric base for the quantitative treatment of the migration dynamics. The inter-regional migration processes in six countries, the Federal Republic of Germany, Canada, France, Israel, Italy and Sweden are briefly discussed under comparative aspects.

In contrast to the previously described models the dynamics may become much more complicated. For instance, if some trend parameters pass critical threshold values (critical values) it may happen, that an originally stable population pattern becomes unstable and a migratory phase transition to a new structure occurs. The agglomeration or clustering of individuals in cities or towns - in other words the formation of urban structure - can be regarded as such a phenomenon.

5.1 Introduction

Migration processes in all their theoretical and empirical aspects are of particular interest to regional science and demography. On the one hand, the underlying motivations for a given kind of migration are relatively well defined and specific and thus available for inquiry (see Courgeau[1,2]). On the other hand, all these motivations must always result in a clear decision to maintain or to change the location in a given interval of time. The number of relocations of a group of individuals per time interval can be counted and are published in the form of migration tables. Understanding the dynamics of these processes in the general framework of our dynamic decision theory is the objective of this chapter. It should be noted that no demographic theory and no reliable forecasting can be made without considering migrations.

A variety of approaches to demographic changes and population movements have been proposed. Classical *demographic* analysis has rarely been oriented to interactions between regions. Rather it is concentrated on the dynamics of *age - class interactions* in one region, closed off from the surrounding world.

The classical Lotka-Keyfitz [3] population model for instance focuses on interactions between age-groups. In the study of populations, however, the assumption of linearity becomes rather unreasonable in any extended time perspective. The exponential growth pattern inherent in most demographic models cannot be more than an approximation, valid for rather limited periods of time. Limits to growth always exist.

For human populations however, the corresponding trend parameters are rather difficult to calculate and to motivate, since social as well as economic interactions have to be considered. Human individuals tend to optimize their decisions.

In the *economy-demography* models (Hotelling,[4] Puu[5]) consumer goods are produced in a production system that can be captured by a production function. The inclusion of production and population leads to highly complicated solution properties, including phase transitions. In a series of publications by Andersson et al,[6-8] and Beckman[9] the public or collective processes have been shown to give rise to nonlinear phenomena as well.

Assuming stationarity Leonardi and Casti[10] derived a logit-like model for the population pattern on the basis of random utility theory. Changes in the nodal public capacity of a region (parks, museums, housing and the like) or in the behaviour of people must lead to changes of the corresponding equilibrium utilities and result in a redistribution of population by migration.

The motivation structure behind migration patterns has been intensely investigated in recent work on the microlevel as well as on the macrolevel. Factors such as the housing market (Clark and Smith[11]), neighbourhood quality, distance from place of work and transportation costs (Clark and Burt[12]), preferences for an urban or rural life style, the image and modernity of a city (*Pumain*[13]), the labour market (Curry[14]) and the structure of the economy of an area (Griffith[15]) have been considered.

However, since in reality most of the migratory systems are not in equilibrium (see Fig.5.5) a *dynamic* framework flexible enough to allow modifications and generalizations is required. Several approaches towards a general theory of migration have recently been developed (see MacKinnon,[16] Sonis[17] and, in particular the articles in Griffith and Lee[18]). The migration model presented here follows the line of argumentation in Haag and Weidlich.[19,20] The application of the model to interregional migration in six countries is described in detail in Weidlich and Haag's book.[21]

5.2 The Stochastic Migration Model

Migration can be formulated as a dynamic choice process by the following argument: let there be L alternatives (the L - regions of the country) to be chosen by an individual. Then the realized population distribution is the result of a sequence of decision processes of all individuals of the population. Therefore, the phenomenon of inter-regional migration fits very well in our dynamic framework of decision processes.

The individual motivations and resulting decisions in the migration process are highly complex. Therefore a reasonable and practicable description of such decisions must be formulated in probabilistic terms. For a member of a certain population there is a certain probability per unit of time of moving from one area to another. These transition rates are formulated in terms of utility functions, which depend on certain trend parameters. When the changes in time of these trend parameters are also known, the dynamics of the system is fully determined. Before going into details, however, we have to draw a general conclusion.

If the individual decisions are stochastic, the evolution of the global system composed of migrating individuals cannot be fully deterministic either. Instead the system must be described by an equation of motion for the evolution of a probability distribution over its possible states.

We shall now consider in detail the migration of one homogeneous population consisting of N members (agents) between L areas. Of particular interest is the macrostate of the migration system characterized by the *population configuration* (this is the now adequate notation for the decision configuration):

$$\boldsymbol{n} = \{n_1, n_2, \ldots, n_L\} \tag{5.1}$$

where n_i is the number of people in region $i = 1, 2, \ldots, L$ and the total population is given by:

$$\sum_{i=1}^{L} n_i = N. \tag{5.2}$$

We seek to understand the dynamics of $\boldsymbol{n}(t)$. To achieve this goal we start from the behaviour of individuals, whose decisions to migrate are governed by comparative considerations of the utility of the origin and destination areas of residence.

5.2.1 The Individual Transition Rates for the Migration Process

In analogy to Chapter 2, we therefore introduce individual transition rates:

$$p_{ij}(\boldsymbol{n}) = \nu_{ij}(t)\exp[u_i(\boldsymbol{n}) - u_j(\boldsymbol{n})] \tag{5.3}$$

composed of a symmetric mobility factor:

$$\nu_{ij} = \nu_{ji} \tag{5.4}$$

and a push/pull factor depending on the utility functions of the origin region $u_j(\mathbf{n})$ and the destination region $u_i(\mathbf{n})$. The utility function $u_j(\mathbf{n})$ can be seen as a measure for the attractiveness of the region j.

Furthermore, it is possible to define a global mobility, $\nu_0(t)$, characterizing the mean mobility of the population under consideration:

$$\nu_0(t) = \frac{1}{L(L-1)} \sum_{\substack{i,j \\ (i \neq j)}}^{L} \nu_{ij}(t), \tag{5.5}$$

However, a multitude of socio-economic factors merges into the utilities $u_j(\mathbf{n})$ and the mobility matrix ν_{ij}. The estimation of the mobilities and utilities and their further analysis is implemented in the following sections of this chapter.

The Mobility Matrix

The mobility matrix $\nu_{ij}(t)$ should include all effects which will either facilitate or impede a transition from j to i independently of any gain of utility. Thus, in particular, the effect of *distance* in its most general meaning will be reflected by the mobility matrix (see also Section 2.1.4). Furthermore, it is not only plausible but also validated in excellent approximation by a regression analysis [21] that $\nu_{ij}(t)$ can be split into two factors:

$$\nu_{ij}(t) = \nu_0(t)\, f_{ij} \tag{5.6}$$

with a time independent *deterrence factor* $f_{ij} = f_{ji}$, describing the effects of *space* (distance) and a time dependent global mobility $\nu_0(t)$. Defining $\nu_0(t)$ as the mean value of the mobility matrix according to (5.5), it follows from (5.5), (5.6) that the f_{ij} must fulfil the normalization condition:

$$\frac{1}{L(L-1)} \sum_{\substack{i,j \\ (i \neq j)}}^{L} f_{ij} = 1 . \tag{5.7}$$

An *effective distance* $D_{ij} = D_{ji}$ comprising all geographic, economic and social distance effects may then be defined by:

$$\exp(-D_{ij}) = \frac{1}{L(L-1)} f_{ij} \tag{5.8}$$

Since, it follows from (5.7), that:

$$0 < \frac{1}{L(L-1)} f_{ij} < 1$$

must hold, the effective distance $D_{ij} = D_{ji}$ is always positive. It is worthwhile to emphasise that only then, if it turns out that the effective distance D_{ij} is highly correlated to the geographical distance d_{ij}, that means, if:

$$D_{ij} \approx \beta\, d_{ij} \tag{5.9}$$

is approximately fulfilled, the mobility matrix assumes the well-known exponential form:

$$\nu_{ij}^{(r)}(t) = L(L-1)\, \nu_{0}(t) \exp(-\beta d_{ij}) . \tag{5.10}$$

It has, however, been shown,[21] that (5.10) is a rather restrictive assumption leading to a considerable reduction of the quality of the total fit.

In some applications the distance matrix d_{ij} is defined as non symmetric: $d_{ij} \neq d_{ji}$. Here, however, we only use a symmetric distance matrix, whereas an eventual non symmetric part of d_{ij} is taken into account by shifts of the utility levels of the regions.[20]

The Utility Function

As mentioned we characterize the attractiveness of a region i for a member of the population by introducing utility functions $u_i(\mathbf{n})$, depending on parameters specific for that region i.

Taking the population numbers as a proxi for size, it can be assumed that regions with a large population attract more people than regions with small population numbers. This population size effect must be distinguished from the size-independent effects, which also influence the utility functions. It is therefore indicated to decompose the total regional utilities:

$$u_i(n_i) = s_i(n_i) + \delta_i(t) \tag{5.11}$$

into a size dependent part, $s_i(n_i)$, and size independent preferences $\delta_i(t)$. Anticipating the result of the analysis to concrete regional systems, it turns out, that the population numbers $n_i(t)$, and the squared population numbers $n_i^2(t)$ are appropriate size-effect variables:

$$s_i(n_i) = \varkappa\, n_i - \sigma\, n_i^2 \,. \tag{5.12}$$

The first term in (5.12) describes agglomeration effects (if $\varkappa > 0$), the second term saturation effects ($\sigma > 0$).

The regional preferences $\delta_i(t)$, describe mainly those parts of the regional attractiveness which depend on socio- economic variables exogeneous to the model dynamics. The corresponding set of keyfactors which in turn are appropriate to describe the temporal changes of the regional preferences $\delta_i(t)$ as well as of the global mobility $\nu_0(t)$ can be found using regression methods.

Since in (5.3) only the difference of the utilities of the origin and destination region appear, all utilities are only defined except for an arbitrary common additive constant. This constant can be adjusted in such a way that the utilities always fulfil the condition:

$$\sum_{i=1}^{L} u_i(t) = 0 . \tag{5.13}$$

For further discussion we introduce the *regional variance* of the utilities:

$$\sigma^2(t) = \frac{1}{L} \sum_{i=1}^{L} u_i(t)^2 \tag{5.14}$$

which can be seen as a measure for the inhomogeneity of the spatial system with respect to their migratory attractiveness.

5.2.2 The Configurational Transition Rates

With the help of the individual transition rates (5.3), it is easy to construct the configurational transition rates, namely the transition rates between different population configurations.

Each of the n_j residents of region j changes to region i with an individual transition rate $p_{ij}(\mathbf{n})$ and thus gives rise to the population configuration transition on the macrolevel:

$$\mathbf{n} = \{n_1, \ldots, n_i, \ldots, n_j, \ldots, n_L\} \rightarrow \mathbf{n}^{(ij)} = \{n_1, \ldots, n_i+1, \ldots, n_j-1, \ldots, n_L\}$$

Hence, the n_j members in region j contribute the term:

$$w_{ij}(\mathbf{n}) = n_j \, p_{ij}(\mathbf{n}) = n_j \, \nu_0(t) f_{ij} \exp[u_i(n_i + 1) - u_j(n_j)] \tag{5.15}$$

to the configurational transition rate $w_t(\mathbf{n})$. Since the transitions between all regions take place simultaneously and independently, the total configurational transition rate is the sum of all contributions (5.15):

$$w_t(\mathbf{n}) = \sum_{i,j=1}^{L} w_{ij}(\mathbf{n}). \tag{5.16}$$

One remark concerning the importance of the term $(n_i + 1)$ in the utility function of (5.15). The factor +1 is of no substantial importance under consideration of the generally high population numbers $n_i \gg 1$ and the uncertainty of the empirical data. However, since the stationary solution is constructed by a product of a great many combinatorical terms, the +1 is of great practical importance from a mathematical point of view (see Section 5.2.4).

Individual birth and death processes in region i induce respective transitions:

$$\mathbf{n} = \{n_1, \ldots, n_i, \ldots, n_j, \ldots, n_L\} \rightarrow \mathbf{n}^{(i+)} = \{n_1, \ldots, n_i+1, \ldots, n_j, \ldots, n_L\}$$

$$\mathbf{n} = \{n_1, \ldots, n_i, \ldots, n_j, \ldots, n_L\} \rightarrow \mathbf{n}^{(i-)} = \{n_1, \ldots, n_i-1, \ldots, n_j, \ldots, n_L\}$$

The corresponding *configurational transition rates* (per unit of time) are denoted as $w_{i+}(\mathbf{n})$, $w_{i-}(\mathbf{n})$, respectively. It is intuitively clear and can be empirically confirmed (Pumain[22]) that these rates can be linked to the individual birth rate $\beta_i(t)$, and the individual death rate $\mu_i(t)$ in region i via:

$$w_{i+}(\mathbf{n}) = \beta_i(t) n_i$$

$$w_{i-}(\mathbf{n}) = \mu_i(t) n_i. \tag{5.17}$$

5.2.3 The Stochastic Equations of Motion

On a stochastic level, fully consistent with the probabilistic description of individual decision processes, we consider in this section the equation of motion for the probability to find a certain population configuration realized at time t. This *configurational probability* :

$$P(\mathbf{n}, t) = P(n_1, n_2, \ldots, n_L, t) \tag{5.18}$$

has to fulfil the normalization condition:

$$\sum_{\mathbf{n}} P(\mathbf{n}, t) = 1 \tag{5.19}$$

where the sum extends over all possible configurations.

If both migratory transitions as well as birth and death processes are taken into account, the configurational probability satisfies the following master equation:

$$\begin{aligned}\frac{dP(\mathbf{n}, t)}{dt} &= \sum_{i,j}^{L} w_{ji}(\mathbf{n}^{(ij)})P(\mathbf{n}^{(ij)}, t) - \sum_{i,j}^{L} w_{ji}(\mathbf{n})P(\mathbf{n}, t)\\ &+ \sum_{i=1}^{L} w_{i+}(\mathbf{n}^{(i+)})P(\mathbf{n}^{(i+)}, t) - \sum_{i=1}^{L} w_{i+}(\mathbf{n})P(\mathbf{n}, t)\\ &+ \sum_{i=1}^{L} w_{i-}(\mathbf{n}^{(i-)})P(\mathbf{n}^{(i-)}, t) - \sum_{i=1}^{L} w_{i-}(\mathbf{n})P(\mathbf{n}, t)\end{aligned} \tag{5.20}$$

In (5.20) we used a more compact formulation since the meaning of the different terms in (5.20) is evident. The birth and death processes are treated in analogy to Section 2.1.3.

5.2.4 The Stationary Solution of the Migratory Master Equation

Subsequently we shall consider the stationary solution $P_{st}(\mathbf{n})$ of the master equation (5.20), which corresponds to a migratory system at equilibrium. The construction is facilitated by the fact, that the transition rates (5.15) satisfy the condition of detailed balance:

$$w_{ji}(\boldsymbol{n}^{(ij)})P_{st}(\boldsymbol{n}^{(ij)}) = w_{ij}(\boldsymbol{n})P_{st}(\boldsymbol{n}), \tag{5.21}$$

which means, that the stationary probability flux from $\boldsymbol{n}$ to $\boldsymbol{n}^{(ij)}$ is equal to the inverse flux from $\boldsymbol{n}^{(ij)}$ to $\boldsymbol{n}$. The repeated application of (5.21) then leads to the following explicit result (details are presented in the Appendix, Section 9.6).

$$P_{st}(\boldsymbol{n}) = \frac{Z^{-1}\,\delta(\sum_{i=1}^{L} n_i - N)}{n_1!\, n_2! \ldots n_L!} \exp[\,2 \sum_{i=1}^{L} F_i(n_i)] \tag{5.22}$$

where

$$F_i(n_i) = \sum_{m=1}^{n_i} u_i(m)\,; \qquad F_i(0) = 0 \tag{5.23}$$

and

$$\delta(\sum_{i=1}^{L} n_i - N) = \begin{cases} 1\,, & \text{for } \sum_{i=1}^{L} n_i = N \\ 0\,, & \text{otherwise} \end{cases} \tag{5.24}$$

and where the factor Z follows from the normalization condition for the probabilities (5.19).

Using *Stirling's* formula for the factorials, the stationary distribution can be written in the form:

$$P_{st}(\boldsymbol{n}) = \frac{\delta(\sum_{i=1}^{L} n_i - N)}{Z} \exp[\sum_{i=1}^{L} \Phi_i(n_i)] \tag{5.25}$$

with

$$\Phi_i(n_i) = 2\,F_i(n_i) - n_i\,(\ln n_i - 1). \tag{5.26}$$

The maxima and minima $\{\hat{n}_1, \hat{n}_2, \ldots, \hat{n}_L\}$ of $P_{st}(\boldsymbol{n})$ can now be determined. They describe the equilibrium configuration (configurations) of highest probability. The extrema of $P_{st}(\boldsymbol{n})$ correspond to the extrema of $\sum_i \Phi_i$ under the constraint $\sum_i n_i = N$. Hence, $\hat{\boldsymbol{n}}$ is found from

$$\delta \left\{ \sum_{i=1}^{L} \Phi_i(n_i) - \lambda \left(\sum_{i=1}^{L} n_i - N \right) \right\} = \sum_{i=1}^{L} \delta n_i \left[2 \frac{\partial F_i(n_i)}{\partial n_i} - \ln n_i - \lambda \right] = 0 \tag{5.27}$$

The *Lagrangian* parameter λ takes into account the constraint (5.2). Since n is a quasi-continuous variable, the sum in (5.23) can be replaced by an integral which amounts to:

$$\frac{\partial F_i(n_i)}{\partial n_i} \approx u_i(n_i) \tag{5.28}$$

Inserting (5.28) into (5.27) and solving for $\hat{\boldsymbol{n}}$ yields:

$$\hat{n}_i = \frac{N \exp[2u_i(\hat{n}_i)]}{\sum_{j=1}^{L} \exp[2u_j(\hat{n}_j)]}, \quad \text{for } i = 1, 2, \ldots, L \tag{5.29}$$

If this set of transcendental equations has one solution only, the distribution is unimodal, and the equilibrium is unique. However, if more than one solution exists, the distribution is multimodal with peaks corresponding to different possible equilibrium states of the migratory system. It depends on the numerical values of the trend parameters in the utility functions, which of these cases is realized.

We shall now discuss the relation between the stationary solution $P_{st}(\boldsymbol{n})$ and the time-dependent solution $P(\boldsymbol{n}, t)$. By definition, the stationary solution is the time-independent solution of the master equation for constant trend parameters. Furthermore, it can be proved (Stratonovich[24]) that *any* time dependent solution approaches the stationary solution for $t \to \infty$. However, in general the migratory system will not be in its equilibrium state. Two cases of *non-equilibrium* should be distinguished.

In the first case the trend parameters are time independent, but the distribution $P(\boldsymbol{n}, t)$ has not yet reached its equilibrium state $P_{st}(\boldsymbol{n})$, so that the migration flows are still time-dependent and the probability distribution approaches the stationary distribution $P_{st}(\boldsymbol{n})$.

In the second case the trend parameters themselves can be slowly time dependent. The transition rates w_{ij} now also become time dependent via their dependence on trend parameters. The master equation (5.19) is still valid in this more general case, but in general the probability distribution $P(\boldsymbol{n}, t)$ will not reach a stationary state at all (except for special cases, for instance when the trend parameters asymptotically approach constant values). Even in the second case we can define a *virtual equilibrium solution* as the stationary solution of the master equation belonging to the momentary set of trend parameters.

The migratory pattern would approach this virtual equilibrium solution if from this time on we would keep the trend parameters constant. Comparing the actually realized population pattern $\boldsymbol{n}^e(t)$ with the virtual equilibrium pattern $\hat{\boldsymbol{n}}$ we can define a *distance from equilibrium* under the momentary trend situation. As a quantitative measure for this *distance from equilibrium*, we introduce the *migratory stress*:

$$s(\boldsymbol{n}^e, \hat{\boldsymbol{n}}) = \frac{1}{2}\,(1 - r(\boldsymbol{n}^e, \hat{\boldsymbol{n}})) \quad \text{with} \quad 0 \le s(\boldsymbol{n}^e, \hat{\boldsymbol{n}}) \le 1 \tag{5.30}$$

where $r(\boldsymbol{n}^e, \hat{\boldsymbol{n}})$ is the correlation coefficient. In Section 5.3.5 the migratory stress for the six countries under consideration is plotted.

5.2.5 Quasi-Deterministic Equations of Motion

As in the proceeding sections we now derive equations of motion for mean values $\bar{\boldsymbol{n}}(t)$ of the regional population pattern. The mean value equations are often already sufficient for comparison with empiric data.

We skip the derivation of the mean value equations here, since the procedure is straightforeward, in analogy to Section 2.2.3. After insertion of the explicitly given transition rates (5.15) we finally obtain:

$$\frac{d\bar{n}_k}{dt} = \sum_{i=1}^{L} \bar{n}_i \nu_{ki} \exp(u_k - u_i) - \sum_{i=1}^{L} \bar{n}_k \nu_{ik} \exp(u_i - u_k)$$

$$+ (\beta_k - \mu_k)\bar{n}_k , \qquad k = 1, 2, \ldots, L \tag{5.31}$$

Evidently, (5.31) is a set of L coupled nonlinear first order differential equations for the mean values of the regional population size. Howver, sometimes it can be justified from empirical considerations that the rate of natural increase:

$$\rho_k(t) \equiv (\beta_k - \mu_k) \approx \rho(t) \tag{5.32}$$

does not depend on the region (sometimes this assumption has to be made because of lack of data). Then it is easy to separate the birth/death processes from the migratory redistribution of population by introducing the relative population shares through:

$$x_k(t) = \frac{\bar{n}_k(t)}{N(t)} \qquad \text{with} \qquad 0 \leq x_k(t) \leq 1 \tag{5.33}$$

Making use of the equation for the total population growth:

$$\frac{dN}{dt} = \rho(t)N(t) \tag{5.34}$$

which can be derived from (5.31), one obtains an equation of motion for the relative population shares $x_k(t)$:

$$\frac{dx_k}{dt} = \sum_{k=1}^{L} x_i \nu_{ki} \exp(u_k - u_i) - \sum_{k=1}^{L} x_k \nu_{ik} \exp(u_i - u_k)$$

$$k = 1, 2, \ldots, L \tag{5.35}$$

The relative dynamics of the spatial population pattern can be simulated by solving (5.35).

5.2.6 The Stationary Solution of the Quasi-Deterministic Equations

If the mobility matrix ν_{ik} and the dynamic utilities u_i are constant with time, the *stationary population shares* x_k^{st} of (5.35) can immediately be obtained and verified by inspection:

$$x_k^{st} = C \exp[2u_k(x_k^{st})], \tag{5.36}$$

where the constant C is given by (5.33):

$$\sum_{k=1}^{L} x_k^{st} = 1 . \tag{5.37}$$

Comparing (5.36, 37) with (5.29) we recognize that the stationary state (states) x_k^{st} of the approximate mean value equations (deterministic equations) coincide with the maximum (maxima) $\hat{n}_k$ of the unimodal (multimodal) stationary probability distribution. This confirms the relation between the fully stochastic and the mean value approach.

In the further discussion of this chapter we make use of the mean value equations (5.31, 35) only. Some general remarks about the structure of the solutions of these equations are due.

We shall first consider the case of constant trend parameters. This means that the ν_{ki} are constants and the utility functions u_i depend on time only via their dependence on the $\boldsymbol{x}(t)$. As we have seen, the stationary solution of (5.35) then fulfils (5.36). Some of these stationary points may be stable, others may be unstable. This mean, that any time dependent solution in the vicinity of a *stable* stationary point (*focus*) will approach this *focus*, whereas at least some time dependent solution in the vicinity of an *unstable* focus will diverge from this focus. Globally, around each stable focus there exists a domain of points (a *basin*) so that all solutions of (5.35) starting from any point of the basin will approach the focus. Therefore, a stable focus is denoted as *attractor.* If no other, more complicated attractors exist (e.g., *limit cycles* or even *strange attractors*), the whole domain of points corresponding to all possible population configurations must decompose into subsets, each being the basin of one of the stable foci. This means, that any time dependent solution of (5.35), wherever it starts will sooner or later be attracted by a stable focus and will finally end up in this equilibrium state.

Second we consider the case of slowly time dependent trend parameters in the utility functions. As a consequence, the situation of the foci will move slowly, too. Normally, however, the solutions of (5.36) will not change their global character to a large extent. The foci will undergo small shifts and the solution will approach the slowly moving stable foci. In certain situations however, this picture changes dramatically. If someone of the trend parameters passes a critical threshold it may occur that an originally stable focus becomes unstable. If the migratory system was in that originally stable focus, it will have to leave this now unstable focus and make a transition towards another now stable focus. We denote this phenomenon as *migratory phase transition* because of its structural analogy to phase transitions in physics.

In Chapter 3 we have treated the simplest model of such a phase transition. In the present chapter it turns out that migratory phase transitions are a general phenomenon to be expected in migration dynamics.

In general the explicit solution of (5.35) can only be found by using different solution algorithms . In simple cases, however, a further analytical investigation is possible. For instance one finds (Weidlich and Haag[25]) for the choice of trend parameters:

$$\nu_{ki} = \nu \; ; \qquad u_i(n_i) = \varkappa\, n_i \; ; \qquad k = 1, 2, \ldots, L \tag{5.38}$$

that the equipartition point (homogeneous distribution of people over the spatial system):

$$\bar{\boldsymbol{n}}_0^{st} = \{\bar{n}, \bar{n}, \ldots, \bar{n}\}, \qquad \text{with } \bar{n} = N / L \tag{5.39}$$

is the only stable focus of the system, as long as $\varkappa$ is below a critical value $\varkappa_c : \varkappa < \varkappa_c$. If, however, $\varkappa$ exceeds $\varkappa_c = L/(2N)$, it can be proved, by linear and nonlinear analysis that the equipartition point becomes unstable and that the only stable foci are now:

$$\bar{\boldsymbol{n}}_i^{st} = \{\bar{n}_1 = n_-, \ldots, \bar{n}_{i-1} = n_-, \bar{n}_i = n_+, \bar{n}_{i+1} = n_-, \ldots, \bar{n}_L = n_-\}$$

$$\text{with } n_+ > n_- ; \qquad i = 1, 2, \ldots, L \tag{5.40}$$

The new stable foci describe - instead of equipartition of population over the regions - the clustering of people in region i. The other $(L - 1)$ regions $j \neq i$ are depleted, but will not be completely empty. Which one of the L regions will gain importance in this simple model depends on initial population density fluctuations.

However, in more realistic situations saturation effects as well as the influence of distance have to be taken into account. Then, via computer simulation the arisal, competition and selforganisation of urban structure can be simulated and interpreted as a migratory phase transition.

Spatial urban economy as well as sociology and psychology are required in order to understand the dynamics of the trend parameters.

5.2.7 Determination of Utilities and Mobilities from Empirical Data

In analogy to the preceding chapters we are now determining the unknown trend parameters in the transition rates w_{ij}. We subdivide the analysis into two main steps:

a) Determination of Utilities and Mobilities

In the first step all trend parameters of our theory which can be directly linked to the migratory process, namely the mobility matrix ν_{ij} and the regional utilities u_i are estimated from empirical data.

b) Selection of Key - Factors

In the second step the trend parameters obtained via step a) are correlated to an appropriate set of socio-economic variables, in order to select a few but relevant *key-factors* for the urban system.

The subdivision of the analysis into these two steps proves to be very useful, since the regional utilities and the mobilities have got a meaning by themselves. On the other hand assumptions concerning the functional dependence of the utilities on certain variables can easily be tested. Mostly we can dispense with a third step, namely a regression analysis of the whole set of key-factors (selected according to step b), since in the case of migration only slight improvements have been observed.

Step a)

We assume that the following empirical (index e) data set is available for a sequence of $t = 1, 2, \ldots, T$ years:

$$\{ n_i^e(t), w_{ij}^e(t)\} \,, \qquad \text{for} \qquad t = 1, 2, \ldots, T$$
$$\text{and} \qquad i, j = 1, 2, \ldots, L \tag{5.41}$$

The migration matrix (5.15) now has to be matched to the empirical migration flows $w_{ij}^e(t)$, by an optimal estimation of the mobilities $\nu_{ij}(t)$ and utilities u_i, $i, j = 1, 2, \ldots, L$.

We use a log-linear estimation (compare Section 2.3.1) in order to obtain explicit expressions for the optimal utilities and mobilities.

The minimization of the functional (5.42):

$$F[\nu,\boldsymbol{u}] = \sum_{t=1}^{T} \sum_{k,l=1}^{L} \Big[\ln w_{kl}^e(t) - \ln[n_l^e(t)\nu_{kl}(t)\exp(u_k(t) - u_l(t)] \Big]^2 \tag{5.42}$$

enables us to calculate in a straightforward manner the optimal utilities and mobilities, if $u_k(t)$ and $\nu_{kl}(t)$ are considered as fitting parameters at disposal.

Since, in general the empirical data set on inter-regional migration is of a good quality the estimation of the trend parameters can be performed yearly without introducing a smoothing procedure as in Section 3.2.6.

The result for the optimal regional utilities is:

$$u_i(t) = \frac{1}{2L} \sum_{k=1}^{L} \ln\left[\frac{p_{ik}^e(t)}{p_{ki}^e(t)}\right], \qquad \text{for } i = 1, 2, \ldots, L \text{ and } t = 1, 2, \ldots, T \tag{5.43}$$

where we have introduced the empiric individual transition rates:

$$p_{ik}^e(t) = w_{ik}^e(t)/n_k^e(t) \tag{5.44}$$

for convenience. The elements $\nu_{ij}(t)$ of the transition matrix are found to read:

$$\nu_{ij}(t) = \nu_{ji}(t) = \sqrt{p_{ji}^e(t) p_{ij}^e(t)} > 0 \qquad \text{for } i, j = 1, 2, \ldots, L \text{ and } t = 1, 2, \ldots, T \tag{5.45}$$

Formulas (5.43) and (5.45) express the utilities and mobilities in terms of the empirically accessible individual transition rates (5.44).

So far we have treated the trendparameters $u_i(t)$ and $\nu_{ij}(t)$ as not depending on further additional parameters. If, however,

$$u_i(t) = u_i(\Gamma_\beta(t)) \qquad \beta = 1, 2, \ldots, B$$

$$\nu_{ij}(t) = \nu_{ij}(\Lambda_\alpha(t)) \qquad \alpha = 1, 2, \ldots, A \tag{5.46}$$

where $\Gamma_\beta(t)$, $\Lambda_\alpha(t)$ can be treated as independent parameters, the minimization can be performed, too. Of particular interest is the estimation of the deterrence factors f_{ij} and the global mobility $\nu_0(t)$ assuming the relation

(5.4). The optimization procedure then yields for the global mobility:

$$\nu_0(t) = \prod_{\substack{k,l \\ k \neq l}}^{L} \left(\frac{\nu_{kl}(t)}{f_{kl}} \right)^{\frac{1}{L(L-1)}} \tag{5.47}$$

and for the deterrence factor:

$$f_{ij} = C \prod_{t=1}^{T} \nu_{ij}(t)^{\frac{1}{2T}} \tag{5.48}$$

where C is to be determined by constraint (5.7) and we have to insert the estimated mobility matrix $\nu_{ij}(t)$ in formulas (5.47,48).

The interpretations of (5.47) and (5.48) are plausible: apart from normalization factors it turns out, that the time-independent deterrence factor f_{ij} is given by the temporal geometric mean value of the mobility matrix $\nu_{ij}(t)$, whereas the spatially aggregated global mobility $\nu_0(t)$ is proportional to the regional geometric mean value. In spite of this, according to definition (5.5) an arithmetic mean value is used for $\nu_0(t)$. Both expressions (5.5) as well as (5.46), however, yield satisfactory results. The reason for the difference in the estimation is based on the more restrictive assumption (5.6), compared with (5.4).

It should be mentioned that instead of the log-linear also a nonlinear estimation can be implemented based on the minimization of expression:

$$F[\nu,u] = \sum_{t=1}^{T} \sum_{k,l=1}^{L} \left[w^e_{kl}(t) - [n^e_l(t)\nu_{kl}(t)\exp(u_k(t) - u_l(t))] \right]^2 \tag{5.49}$$

The results of this optimization are slightly better, but cannot be written in analytical form. However, (5.49) is well defined as an optimization problem and different methods of its numerical treatment as can be used.

A direct survey of the quality of the fitting procedure can be given by plotting the fitted *theoretical* transition rates $w^{th}_{ij}(t)$ versus the empirical rates $w^e_{ij}(t)$. Figures 5.1 and 5.2 show the fit in the case of the Federal

Republic of Germany for the log-linear and nonlinear estimation (L = 11 regions, T = 27 years of registration, and 2,970 elements w_{ij}). Some comparative information about the estimations including tests of significance are summarized in Table 5. 1.

Table 5.1

Comparison of six versions of estimation in the case of the Federal Republic of Germany

	log-linear estimation			nonlinear estimation		
Figure	(5.1 a)	(5.1 b)	——	(5.2 a)	(5.2 b)	——
form of the mobility	(5.4)	(5.6)	(5.10)	(5.4)	(5.6)	(5.10)
number of parameters	1,782	379	325	1,782	379	325
R^2	0.9973	0.9798	0.7895	0.9992	0.9812	0.7895
$\bar{R}^2$	0.9933	0.9768	0.7636	0.9981	0.9784	0.7636
F	248.1	331.0	30.5	882.4	355.7	30.5

As expected the nonlinear estimation leads to slightly better results in all cases than the log-linear estimation procedure. The transition from general mobilities (5.4) to the reduced mobilities (5.6) leads to an increase of deviations from the 45°-line. On the other hand, the results are still quite satisfactory. The use of (5.10) instead of (5.4), (5.6) provides a much worse estimation, although the number of fitting parameters is only slightly reduced.

Step b)

The estimation procedure, described in step a) yields with some accuracy the time series for the trend parameters $\boldsymbol{\tau}(t) \in \{\nu_0(t),\ u_i(t)\}$. These trend

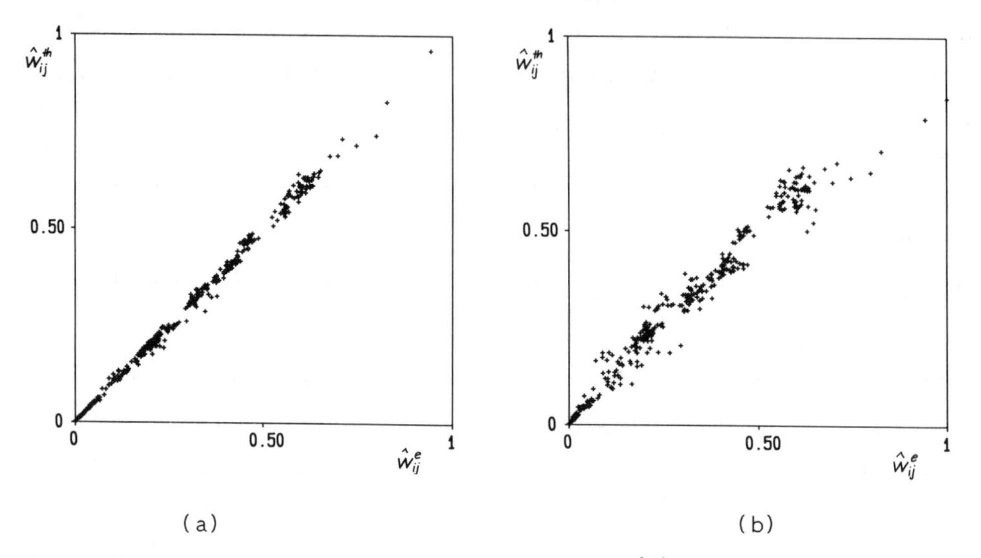

Figure 5.1 : Scaled theoretical migration matrices $\hat{w}_{ji}^{th}(t)$ plotted versus scaled empirical matrices $w_{ji}^{e}(t)$ for the log-linear estimation

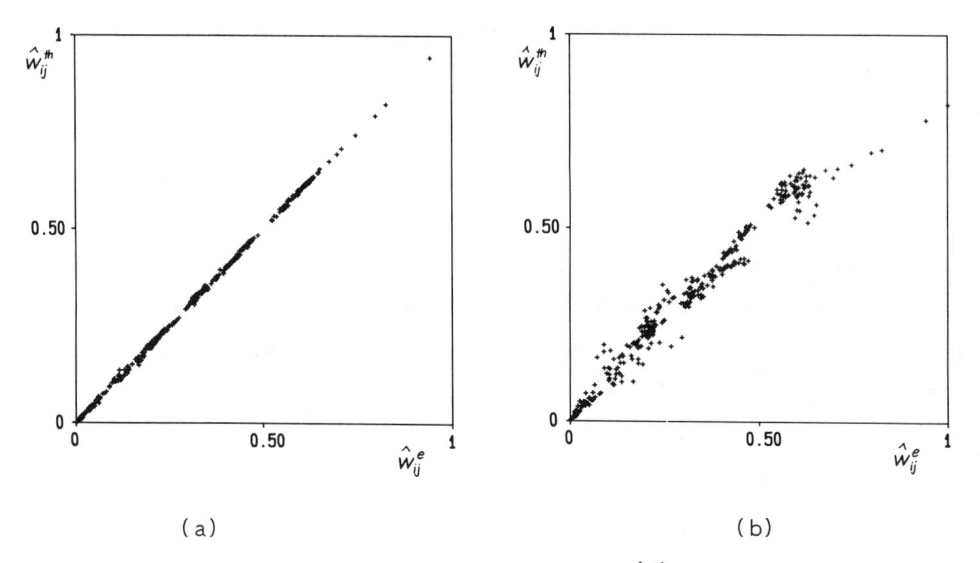

Figure 5.2 : Scaled theoretical migration matrices $\hat{w}_{ji}^{th}(t)$ plotted versus scaled empirical matrices $w_{ji}^{e}(t)$ for the nonlinear estimation

parameters in their turn determine the dynamics of the migration system. Therefore, it is promising to correlate the trend parameters $\tau(t)$ to a set of exogeneous socio-economic variables $\Omega_\alpha(t)$, $\alpha = 1, 2, \ldots, A$; $t = 1, 2, \ldots, T$. These sets of trend parameters are properly standardized and detrended, in the time interval $1 \le t \le T$.

As described in Section 2.3.3 we prefer the ranking regression algorithm (Reiner and Munz[23]) which is based on the following ideas: let us assume that a trend parameter can be represented by a linear combination of explanatory variables (socio-economic variables). In general, however, socio-economic variables can be highly correlated to each other (e.g., the number of households in a region and the number of people living in that region). Therefore, it seems to be adequate to orthogonalize the data base and to select the socio-economic variables with respect to their relevance. Time delays have to be considered, too. The finally chosen socio-economic variables are then denoted as *key-factors*.

5.3 Comparative Analysis of Inter-Regional Migration

The inter-regional migration processes within Federal Republic of Germany, Canada, France, Israel, Italy and Sweden have been described and discussed in detail in Weidlich and Haag[21] under comparative aspects. Therefore, only a brief survey is given in this section.

Before beginning with the comparative analysis it is worthwhile to remember that inter-regional migration is only one but substantial share of the total migration and that migration is only one part of the total evolution of the population. The (registered) total volume of migration consists of intra-urban migration, inter-urban migration between cities belonging to the same region, and inter-regional migration. For example, the relative share of inter-regional migration versus the total volume of migration in the Federal Republic of Germany is about 25 per cent (1983). The order of magnitude of this relative share is about the same in all countries under consideration.

It is also remarkable, that the trends of intra-regional and inter-regional migration need not coincide. In the Federal Republic of Germany, for instance, one observes (1950-1987) a slow decrease of inter-regional migration but on the other hand also a slow increase of intra-regional migration. Thus, in the course of time short distance adaptations seem to be preferred over long distance moves.

The assumption of a space-independent rate of natural increase $\rho(t)$, turns out to be only a rough approximation in some countries, especially in countries like Canada and Israel with waves of immigrants. As a consequence of this nonseparability of migration and birth/death processes forecasting requires the treatment of the full set of equations (5.31).

On the other hand, the regression analysis for the trend parameters $\nu_0(t)$, f_{ij}, $u_i(t)$ also remains valid in this more general case. But the regional preferences and the mobility now also reflect the effect of regionally inhomogeneous immigration, emigration or birth/death rates.

In order to keep inter-regional migration as a relevant factor of population evolution, the regions chosen should not be *too large.* In particular, they should be chosen *as homogeneous as possible* in economic, ethnographic, climatic and other aspects. Indeed, their description in terms of regional utilities becomes more meaningful and interpretable if socio-economic key-factors can be attributed to one region as a whole, and not to disparate subunits. In this case an individual will consider a region more or less as one unit with respect to his migration decisions.

On the other hand, the regions should not be *too small,* since the model is based on mean value equations and the regional populations and the migration flows should not be too small numbers. A further practical reason for not too small regions is the availability of spatially disaggregated data.

Boundaries should not divide cities, since inter-regional migration then would be biased by short distance effects.

5.3.1 Choice of Comparable Socio-Economic Variables

Step b) in our analysis consists of representing the global mobility and the utilities in terms of socio-economic key-factors. Unfortunately the data sets available in different countries do not coincide. However, one can identify classes of variables, so that the variables available for every country belong to one of the following classes:

Size and population structure
Labour market
Housing market
Industry, investments
Public sector, politics
Living standard
Climate

The results of the comparative analysis presented here primarily focus on the question which class of variables - and not which specific variable - proves to be most important.

5.3.2 The Global Mobility under Comparative Aspects

In Figures 5.3 a - f the temporal evolution of the global mobility for the six countries is represented. The mobility is scaled on the temporal mean value 1 in order to avoid effects depending on the chosen regional subdivision of the country.

In Sweden, Italy and the Federal Republic of Germany a permanent long term decrease of the global mobility can be observed, beginning in 1960 in the Federal Republic of Germany and after 1970 in Sweden and Italy. In France, however, the mobility increases until 1975 and begins to decline afterwards. Regional development programms may have led to this trend differing from the other European countries. Another reason for the prolongation of the increase of the mobility rate in France until 1975 may be, that France comes later, as compared with countries like the U.K., FRG or Sweden, in the urbanization process. Its urbanization rate reaches only 57% in 1954 and 73% in 1975.

On the other hand, in Canada and Israel, whose histories differ characteristically from those of European countries, only a slight decline or even no decline at all of the mobility can be observed. This may also be due to the fact, that the preceding immigration waves afterwards lead to a higher level

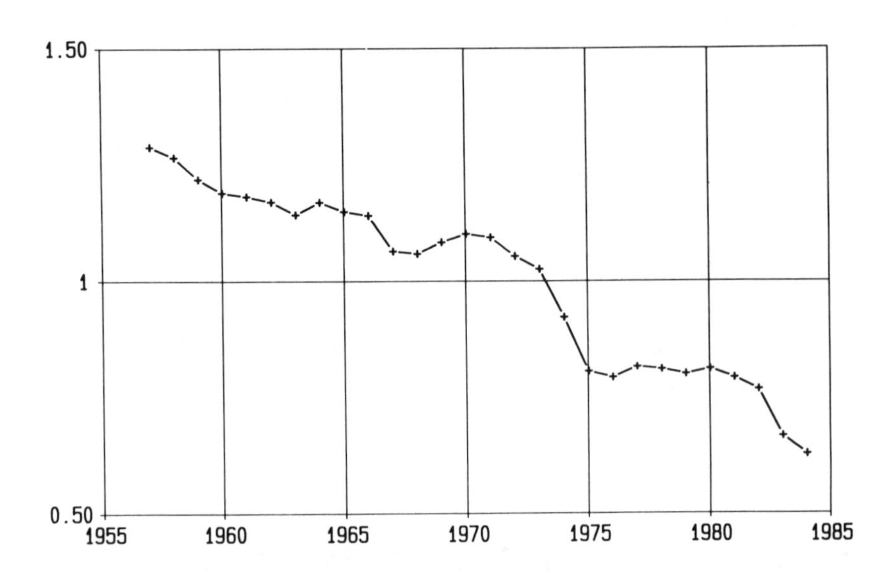

Figure 5.3a: The scaled global mobility $\nu_0(t)$ of the Federal Republic of Germany

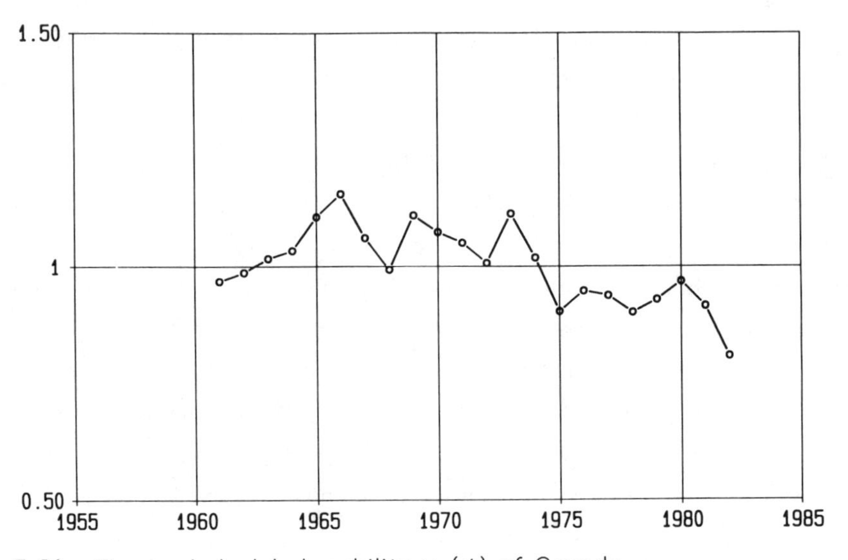

Figure 5.3b: The scaled global mobility $\nu_0(t)$ of Canada

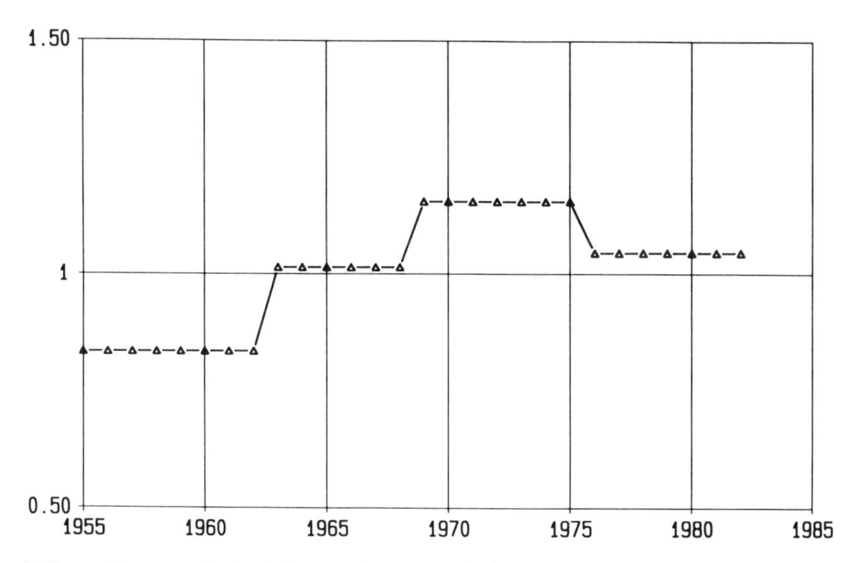

Figure 5.3c : The scaled global mobility $\nu_o(t)$ of France

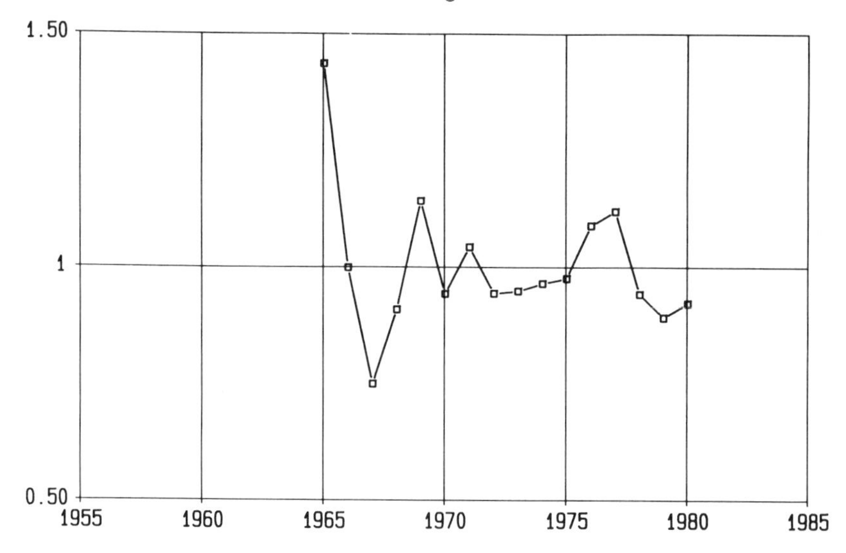

Figure 5.3d : The scaled global mobility $\nu_o(t)$ of Israel

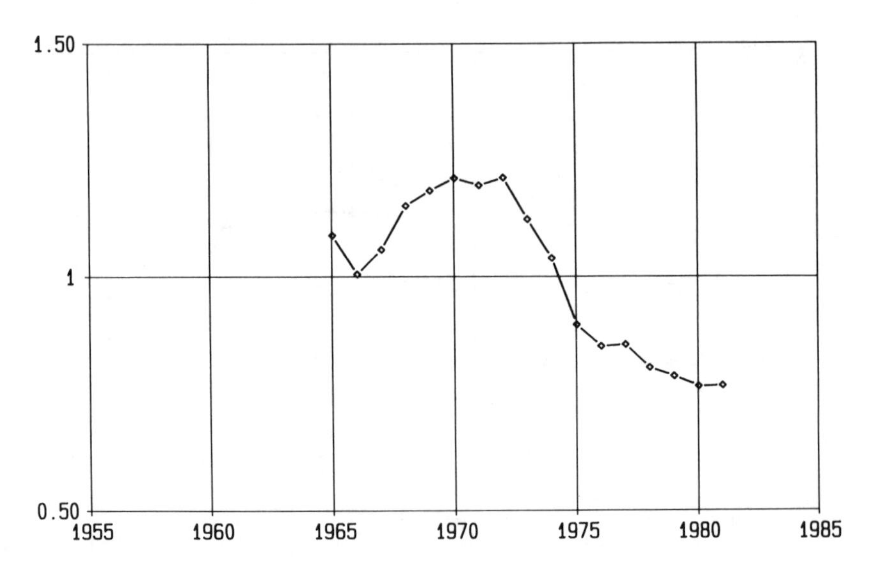

Figure 5.3e: The scaled global mobility $\nu_0(t)$ of Italy

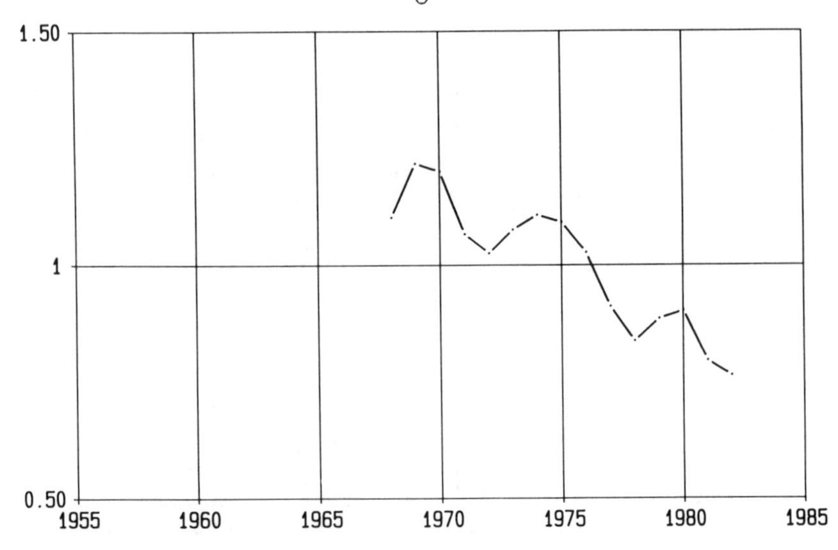

Figure 5.3f : The scaled global mobility $\nu_0(t)$ of Sweden

of inter-regional mobility.

The finer details of the evolution of the global mobility $\nu_0(t)$ superimposing the long term trend are also of interest. Firstly, the economic recession seems to have influenced the global mobility everywhere. Its particularly steep decrease from 1972 to 1975 in the FRG and Italy and its dip in 1972 in Canada and Sweden may confirm this supposition.

Short term business cycles have their effect on the mobility, too. Whenever economic booms can be identified in one, or simultaneously in several countries, as for instance in 1969 in the FRG, the mobility exhibits a synchronous increase.

The key-factors may be assigned to the three classes: labour market, housing market and living standard . Since the correlation of the global mobility to labour market variables are positive in all countries, we can conclude that prosperity induces a higher level and recession a lower level of mobility. The diminution of mobility in a period of recession of course may enhance the negative effects of the latter. It is also easily understandable that housing constructions correlate positively, since the housing market is linked to moving families.

5.3.3 The Regional Utilities and Preferences of the Federal Republic of Germany

The stepwise regression of utilities on socio-economic variables leads to the remarkable result that in all six countries the regional population numbers $n_i(t)$ and its square $n_i^2(t)$, are the two most important key-factors as supposed in (5.12). This means that the decomposition (5.11) of the regional utility function can be justified. Since scaled population numbers (population shares), according to (5.33) are used, the agglomeration parameters $\varkappa$ and the saturation parameters σ of the different countries can immediately be compared (see Table 5.2).

Table 5.2

Agglomeration parameter $\varkappa$ and saturation parameter σ for different countries (t-values in parantheses)

country	$\varkappa$		σ	
Federal Republic of Germany	1.300	(151)	0.258	(27)
Canada	1.546	(105)	0.313	(15)
France	0.831	(90)	0.124	(15)
Israel	1.250	(92)	0.179	(16)
Italy	1.285	(73)	0.245	(14)
Sweden	1.106	(66)	0.295	(5)

The regression analysis yields positive $\varkappa$ and σ - values for all countries which confirm our interpretation. The parameter values of $\varkappa$, σ exceed critical values, so that agglomeration trends prevail and a uniform population distribution becomes unstable. Therefore, the clustering of people in cities can also be understood from this point of view. In other words, there is a preference for high density living, but only up to a certain level.

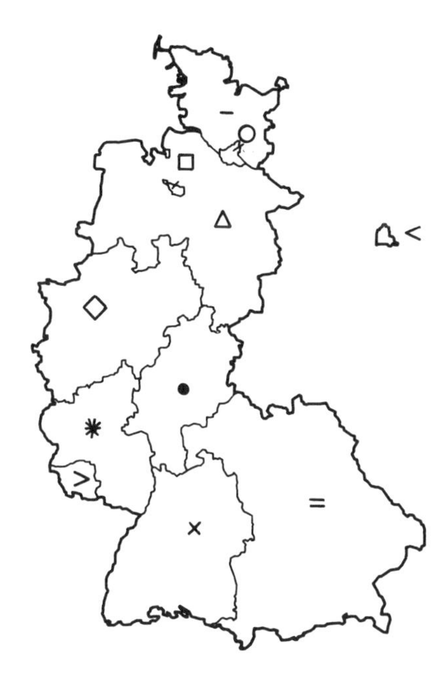

+ Schleswig-Holstein
o Hamburg
Δ Niedersachsen
◻ Bremen
◇ Nordrhein-Westfalen
· Hessen
* Rheinland-Pfalz
x Baden-Württemberg
= Bayern
> Saarland
< Berlin (W)

Figure 5.4: The map of the Federal Republic of Germany and its subdivision into federal states

In Figure 5.5 a, b the regional utilities and preferences for the FRG are shown. The Federal Republic of Germany is divided into 10 federal states and the region of Berlin (W) (see Fig. 5.4). The names and the symbols used in graphical presentations of Figure 5.5 are also listed in Figure 5.4.

The regional utilities are estimated according to (5.49). By definition the utilities reflect the overall attractiveness of a federal state for a migrant. At a first glance, it can be seen that the utilities of all federal states remain relatively stable over a long time period, with the exception of Berlin West (<). The Berlin-effect is mainly due to the flow of refugees from Berlin to the FRG before the erection of the Berlin wall (1961). This leads to a formal underestimation of the *attractiveness* of Berlin in the period from 1957 to 1962, since the simultaneous refugees from the DDR have not been registred as immigrant flow. After the erection of the wall between East and West Berlin in 1961 the situation in Berlin stabilized partially due to economic subsidies by the local administration and the government of the FRG. Since 1982 the differences in utilities is again increasing. The more high-tech oriented regions (=) Bayern, (x) Baden-Württemberg, (•) Hessen, and (<) Berlin gained importance in spite of the coal and steel oriented region (>) Saarland, and the

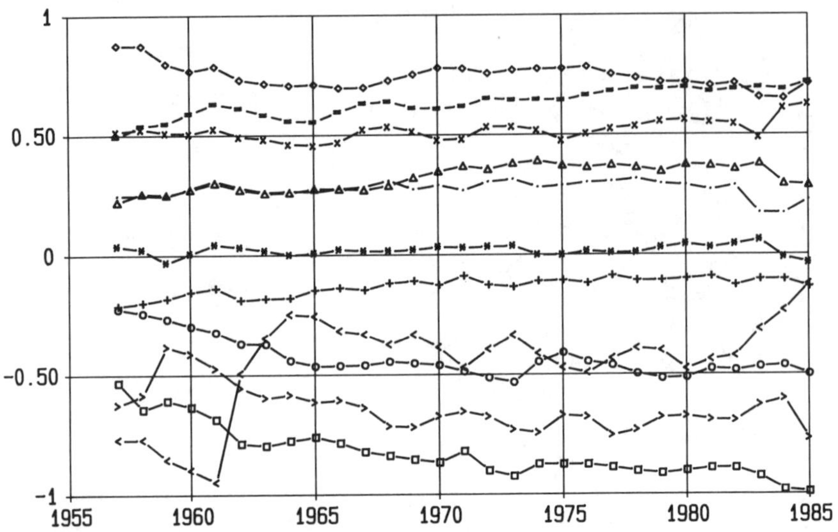

Figure 5.5a: The estimated regional utilities $u_i(t)$ of the FRG

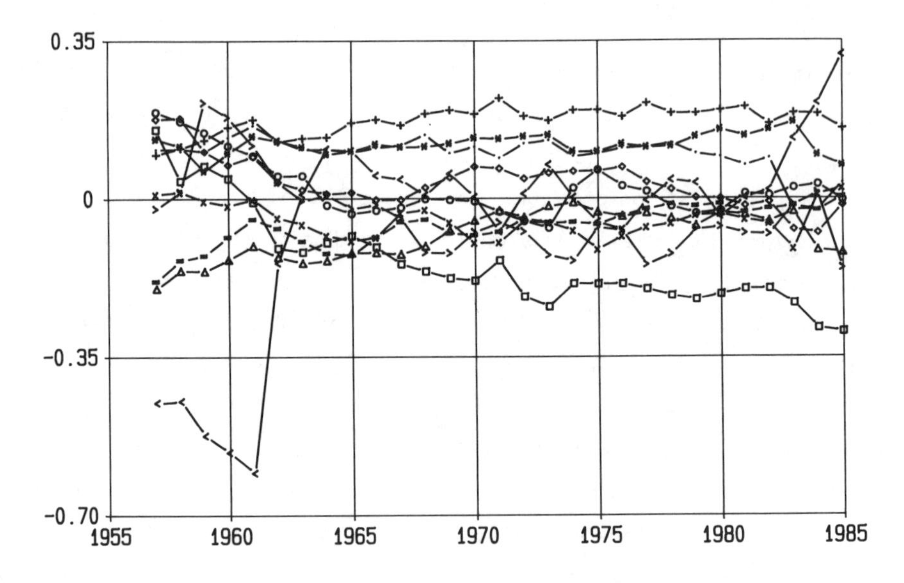

Figure 5.5b: The regional preferences $\delta_i(t)$ of the FRG

international harbours (o) Hamburg, and (□) Bremen. The population rich (◇) Nordrhein-Westfalen seem to have stabilized.

In our key-factor analysis 32 socio-economic variables have been tested. However, only 6 key-factors are necessary to represent the complicated spatio-temporal structure of the regional utilities with some accuracy. The influence of the size effect (x, σ) is still dominant.

The remaining four intensive key-factors are: number of overnight stays per capita, the export structure index, the unemployment rate and the percentage of employment in the tertiary sector. The results of the regression analysis and the considered statistical tests are listed in Table 5.3.

Table 5.3

Ranking regression analysis of the regional utilities of the FRG from 1960 to 1985.

variable	coefficient	time lag (years)	t - value
population number	1.041	0	207.2
square of population	0.181	0	32.7
overnight stays	0.158	0	17.8
export structure index	0.467	2	9.4
unemployment rate	-0.089	0	-6.0
tertiary sector	0.214	0	5.6

correlation R^2: 0.9822

corrected $\bar{R}^2$: 0.9817

F - test value : 2358.4

In a series of publications Birg et al[26-28] and Koch and Gatzweiler[29] assumed as a working hypothesis that the change of location of an individual is subject to:

- the possibility to earn a higher income
- a higher disposal of services, public and private goods
- a better environment and social structure
- a good infrastucture in order to use facilities in other regions of the country

The set of key-factors selected by the ranking regression analysis used here can be attributed to these four different fields of influence.

For a detailed discussion of the different key-factors as well as their interpretation see Weidlich and Haag.[21]

In Figure 5.6 the estimated regional utilities (plotted with symbols according to Figure 5.4) are compared with the regional utilities represented by a linear combination of the six socio-economic key-factors (solid lines) listed in Table 5.3.

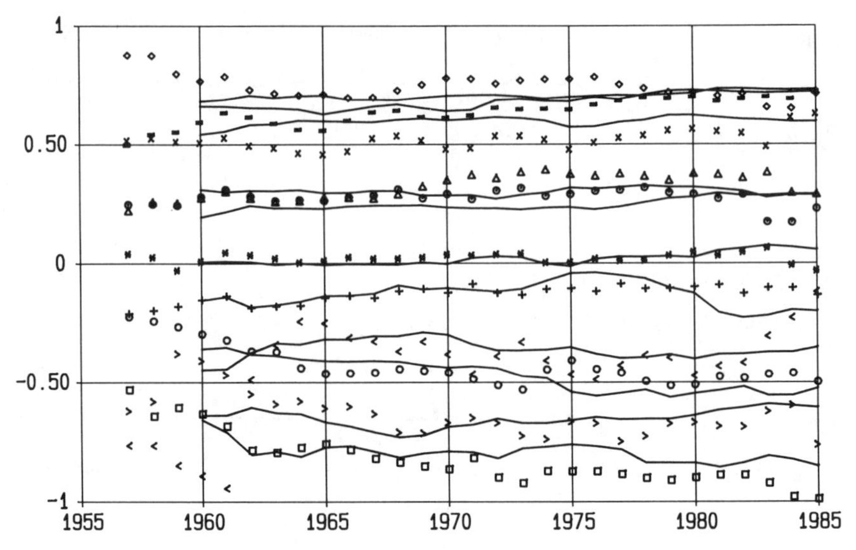

Figure 5.6 : The key-factor analysis of the regional utilities of the FRG

5.3.4 Comparison of the Variance of Utilities

We now focus on the variance of utilities and preferences of the different countries. They present global measures of regional differentiation, where the utilities include and the preferences exclude the population size effect. In Fig. 5.7 a comparative representation of the variance of the utilities is shown.

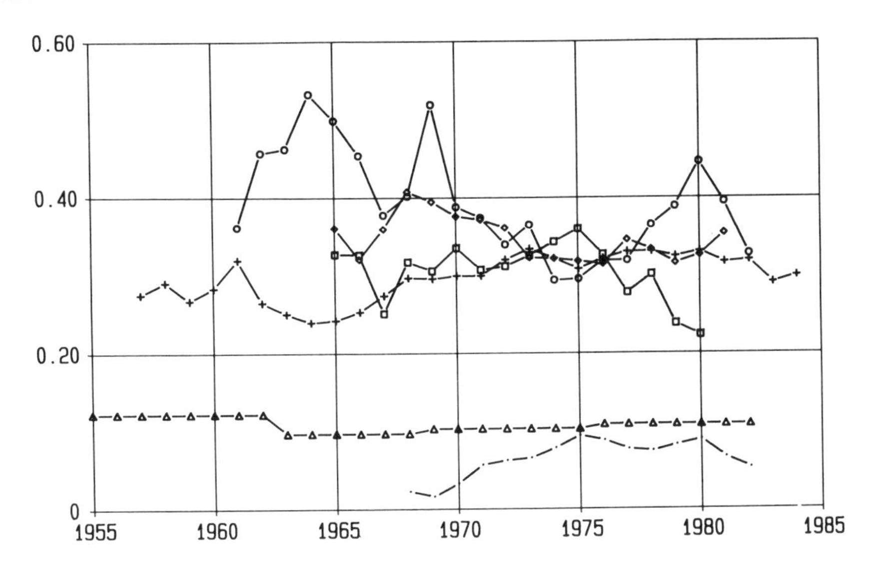

Figure 5.7 : Comparison of the temporal evolutions of the variance of utilities. (+) FRG, (o) Canada, (Δ) France, (□) Israel, (◇) Italy, (•) Sweden

The temporal evolution of the utility variance seems to be of some interest: the variations with time are highest in the case of Canada. The peaks of this variance in 1964, 1969 and 1980 coincide with peaks in the regional preference of Alberta, British Columbia and Ontario, so that the variations could be interpreted by a different pace in economic development of different regions.

In Italy one observes a decline of the utility variance from 1968 to 1975 remaining on this lower level afterwards. This trend to more homogeneity means that migration, perhaps initiated by regional development programms, somewhat smoothed out the differences of regional utilities.

It is remarkable that Israel's utility variance was increasing from 1967 to 1975, but declining afterwards until 1980. Looking at the regional utilities one can see that the later decline is due to an increase of the utility of the non-metropolitan areas in the late 1970s so that afterwards the country becomes more homogeneous.

The Federal Republic of Germany has practically stable regional differentiation with a very small long term increase. This effect is mainly governed by population concentration.

5.3.5 Comparison of the Migratory Stress

We introduced the migratory stress (5.30) as a measure of distance of the population distribution from its (virtual) equilibrium population distribution.

In Fig. 5.8 for comparative purpose the correlation coefficient $r(\mathbf{n}^e, \mathbf{n})$ is depicted. The FRG and Sweden appear to be very close to equilibrium during the period of observation. This is understandable in the case of Sweden with a smooth, undisturbed and balanced development of regions. However, it is notable that in spite of the over 10 million post-war migrants the FRG after 1955 has almost reached the migratory equilibrium. This can be interpreted as an early integration of these refugees into their new regional environment. In this sense the *Akademie für Raumforschung und Landesplanung (Hanover)* had made strong efforts in the early post war period to eleborate plans for the distribution of the new citizens over the country.

Both France and Canada have strong and growing deviations from the virtual equilibrium. In France the correlation coefficient $r(\mathbf{n}^e, \mathbf{n})$ reached a value below 0.8 in 1975. A general reorientation of the migratory trends from *centralization* to a preference of *peripheral* regions may be one of the reasons. Obviously it needs time until the French system reaches its new equilibrium state with respect to the new trends. Canada also exhibits strong fluctuations of the migratory stress. It can be observed that the migratory stress is growing parallel to peaks of the utility variance. The relatively sudden arisal of economic booms in Alberta, British Columbia and Ontario has set the migratory system into stronger deviations from its virtual equilibrium.

Italy reached the maximum of its migratory stress in the late 1960s and showed a slow decrease afterwards. Regional development programms set into effect in the 1960s and initiating new migration trends are probably re-

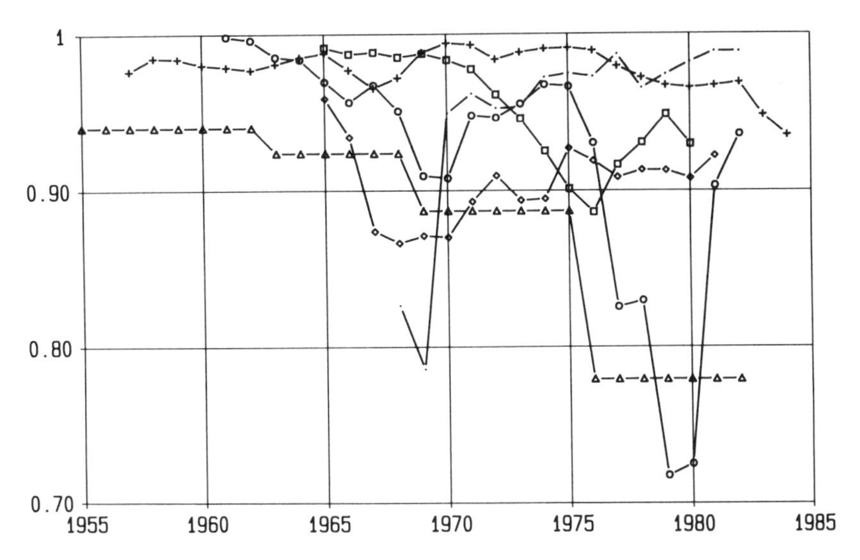

Figure 5.8: Comparison of correlation coefficients $r(n^e, n)$. (+) FRG, (o) Canada, (Δ) France, (□) Israel, (◇) Italy, (·) Sweden

sponsible for this peak of the migratory stress which slowly reduces on the way to equilibrium afterwards.

Israel also starts with almost an equilibrium situation before 1970, until a maximum of deviation from equilibrium was reached in 1976. Afterwards a movement towards equilibrium sets in. This temporal evolution of the migratory stress runs parallel to the evolution of the variance of the utilities.

Chapter 6

Chaotic Evolution of Migratory Systems

On our way from simplicity to complexity we are now approaching the field of interacting agents or subpopulations. We shall see that the inter-group and intra-group interactions decisively influence the choice behaviour of individuals. A phase transition from an ordered state to a chaotic migration pattern may then occur on the macrolevel.

6.1 Introduction

Chaotic motion of the *conventional type* was for a long time a universally well-known phenomenon treated by statistical methods in physics. The prototype of *conventional chaos* is the *Brownian* motion: a dust particle suspended in a gas or fluid executes a random motion because of its irregular collisions with molecules. It is characteristic of *conventional chaos*, that a large number of external microscopic degrees of freedom - the colliding molecules in the example mentioned - leads to random forces generating chaotic motion.

But it came as a great surprise to scientists in the last few decades that another type of chaos is also possible. This new type of chaos is denoted as *deterministic chaos*. It can arise in a set of coupled deterministic differential equations (three or more), or even in systems described by difference equations (one or more).

Although in the case of *deterministic chaos* the trajectory approaches, or remains within, a finite domain of the space of variables, it does not show any periodicity; that means it never exactly returns to the same point in phase space, Correspondingly, the *Fourier* analysis of the chaotic orbit yields a continuous broad band spectrum of frequencies instead of the few selected frequencies which any regular periodic motion would exhibit. Another important characteristic of chaotic motion is the behaviour of the *Lyapunov-exponent*.

In the meantime many examples of deterministic chaos referring to concrete systems have been found and analyzed in natural science. An excellent survey about recent developments in the field of chaotic behaviour is given in Haken's book.[1]

If deterministic chaos appears in simple models of natural science, it should be even more expected that it also happens in complex systems of social science. Therefore the investigation of the chaotic behaviour of economic macrovariables has gained more and more interest. The question of any possible predictability of economic long term development is a crucial task and challenge of economists. Dendrinos[2] and Mosekilde et al[3] have shown that urban migration systems may exhibit strange attractors giving rise to a chaotic flow pattern in the spatial system. We shall demonstrate in this chapter, that deterministic chaos can appear in our general migration model even under less restrictive assumptions.

6.2 The Migratory Master Equation and Mean Value Equations for Interacting Populations

In Chapter 2 we have shown that the aggregation of individual actions - the individual decision process - produces the micro-foundation of the macro-behaviour of the individuals. In Chapter 5 we considered the migration acts of individuals between different regions of a country. We stated as a fundamental assumption that the total population of individuals can be treated as homogeneous with respect to their migratory decision processes. The decision behaviour of individuals to migrate was assumed to be based on a comparison of regional utilities.

We shall now assume that the total population of individuals $N(t)$ consists of $\alpha = 1, 2, \ldots, A$ subpopulations *differing* in their migratory choice behaviour.

This means we skip the assumption of *one* homogeneous population and generalize our model by considering the evolution pattern of subpopulations. The space is devided into L regions. The number of individuals of subpopulation α, living in region $i = 1, 2, \ldots, L$, at time t is $n_{\alpha i}(t)$. For the sake of simplicity we ignore birth and death processes and consider the migration of the subpopulations between the L regions. Hence, the total number N_α of subpopulation α remains constant:

$$\sum_{i=1}^{L} n_{\alpha i}(t) = N_\alpha , \tag{6.1}$$

and the total number of individuals is given by:

$$\sum_{\alpha=1}^{A} N_\alpha = N . \tag{6.2}$$

In analogy to Chapters 2 and 5 we introduce the dynamics in three steps:

1. Definition of individual transition rates (decision rates) under consideration of the whole complex distribution pattern of all individuals
2. Formulation of the dynamic equations of motion for the probability distribution to find a certain migration pattern to be realized
3. Derivation of quasi-deterministic equations for the population numbers from the stochastic system

These three steps are implemented in the following subsections.

6.2.1 Inter-Group and Intra-Group Interactions of Individuals

We introduce individual transition rates $p^\alpha_{ij}(\boldsymbol{n})$ for the transition of one member of subpopulation α from region j to region i per unit of time. The $p^\alpha_{ij}(\boldsymbol{n})$ will depend in general on the whole distribution of all subpopulations

$$\boldsymbol{n} = \{n_{\alpha i}\} , \qquad \text{for } \alpha = 1, 2, \ldots, A; \text{ and } i = 1, 2, \ldots, L \tag{6.3}$$

A sufficiently general and flexible assumption for the individual transition rates is given by (see also Chapter 2)

$$p_{ij}^{\alpha}(\mathbf{n}) = \nu_{ij}^{\alpha} \exp[u_i^{\alpha}(\mathbf{n}) - u_j^{\alpha}(\mathbf{n})], \tag{6.4}$$

where

$$\nu_{ij}^{\alpha} = \nu_{ji}^{\alpha} \tag{6.5}$$

is the symmetric mobility matrix of subpopulation α, which in general will depend on distance (compare Section 5.2.1) between the regions i and j. As a measure for the attractiveness of region i for members of subpopulation α we introduced the *regional utilities* $u_i^{\alpha}(\mathbf{n})$. A plausible form of the regional utilities reads:

$$u_i^{\alpha}(\mathbf{n}) = \delta_i^{\alpha} + \sum_{\beta=1}^{A} \varkappa^{\alpha\beta} n_{\beta i} + \ldots \tag{6.6}$$

where we have written the u_i^{α} as a truncated *Taylor-expansion*.

Obviously, we have the following interpretation: the δ_i^{α} describes the preference of region i to members of subpopulation α. The *intra- group interaction parameter* $\varkappa^{\alpha\alpha}$ describes the agglomeration (clustering) trend of subpopulation α (for $\varkappa^{\alpha\alpha} > 0$). The *inter-group interaction* (interaction between different subpopulations) is represented by the trend parameters $\varkappa^{\alpha\beta}$ (for $\alpha \neq \beta$). If the inter-group interaction parameter fulfils $\varkappa^{\alpha\beta} > 0$, population α prefers to live together with population β, whereas for $\varkappa^{\alpha\beta} < 0$ subpopulation α prefers to separate from subpopulation β, in different regions. Ghetto-formation for example, requires such a negative inter-group interaction (Weidlich and Haag[4]).

The total transition rate for a transition per unit of time of any one member of subpopulation α from region j to i is then:

$$w_{ij}^{\alpha}(\mathbf{n}) = n_{\alpha j}\, p_{ij}^{\alpha}(\mathbf{n}) = \nu_{ij}^{\alpha}\, n_{\alpha j} \exp[u_i^{\alpha}(\mathbf{n}) - u_j^{\alpha}(\mathbf{n})] \tag{6.7}$$

The total transition rates are the basic elements of the stochastic equations of motion.

6.2.2 The Master Equation for Interacting Subpopulations

Choice processes of members of the different subpopulations constitute the basis of the dynamic evolution of the population configuration $\boldsymbol{n}$ (6.3).

We introduce the probability $P(\boldsymbol{n}, t)$ to find the population configuration $\boldsymbol{n}$ at time t. At any time, $P(\boldsymbol{n}, t)$ satisfies the probability normalization conditions :

$$\sum_{\boldsymbol{n}} P(\boldsymbol{n}, t) = 1 , \tag{6.8}$$

where the sum extends over all possible configurations $\boldsymbol{n}$.

A probability balance consideration (see Chapter 2) now yields the fundamental master equation:

$$\frac{dP(\boldsymbol{n},t)}{dt} = \sum_{\alpha=1}^{A} \sum_{i,j=1}^{L} [w_{ji}^{\alpha}(\boldsymbol{n}^{(\alpha,ji)})P(\boldsymbol{n}^{(\alpha,ji)},t) - w_{ij}^{\alpha}(\boldsymbol{n})P(\boldsymbol{n},t)] \tag{6.9}$$

Here, the configuration $\boldsymbol{n}^{(\alpha,ji)}$ arises from the configuration $\boldsymbol{n}$ by the substitution $n_{\alpha i} \rightarrow (n_{\alpha i} + 1)$, $n_{\alpha j} \rightarrow (n_{\alpha j} - 1)$, whereas all other $n_{\beta k}$ remain unchanged.

6.2.3 The Deterministic Equations for Interacting Subpopulations

It is easy to derive from (6.9) quasi-closed equations of motion for the time-dependent mean values:

$$\overline{n_{\alpha i}(t)} = \sum_{\boldsymbol{n}} n_{\alpha i} P(\boldsymbol{n},t) \tag{6.10}$$

according to the procedure described in Chapter 2. The approximate mean value equations for the spatial distribution pattern of members of the subpopulation α then reads:

$$\frac{d\overline{n_{\alpha i}(t)}}{dt} = \sum_{j=1}^{L} \bar{n}_{\alpha j}(t)\nu_{ij}^{\alpha}\exp[u_i^{\alpha}(\bar{\boldsymbol{n}}) - u_j^{\alpha}(\bar{\boldsymbol{n}})] + \sum_{j=1}^{L} \bar{n}_{\alpha i}(t)\nu_{ji}^{\alpha}\exp[u_j^{\alpha}(\bar{\boldsymbol{n}}) - u_i^{\alpha}(\bar{\boldsymbol{n}})] \qquad (6.11)$$

for $\alpha = 1, 2, \ldots, A$ and $i = 1, 2, \ldots, L$.

6.2.4 The Exact Stationary Solution of the Deterministic Equations

Our further study of the migration of interacting subpopulations only refers to the meanvalue equations (6.11). For a given set of trend parameters δ_i^{α} and $\varkappa^{\alpha\beta}$ the equations (6.11) represent a system of autonomous nonlinear differential equations for the population numbers $\overline{n_{\alpha i}(t)}$.

It can easily be seen that the stationary solution of (6.11) obeys the relation:

$$n_{\alpha i}^{st} = C^{\alpha}\exp[2u_i^{\alpha}(\boldsymbol{n}^{st})], \qquad (6.12)$$

with a normalization constant C^{α} which has to be determined for each subpopulation separately:

$$C^{\alpha} = \frac{N_{\alpha}}{\sum_k^L \exp[2u_k^{\alpha}]} . \qquad (6.13)$$

The stationary population pattern $\boldsymbol{n}^{st}$ therefore, can be obtained by solving the coupled system of transcendental equations (6.12). However, in general this will be a nontrivial problem.

It is now possible to classify the solutions of (6.11) according to the number of subpopulations A, and the number of regions L under consideration. Because of the constraint (6.1) the number of independent variables is given by $A(L-1)$.

In Chapter 5 and Haag and Weidlich[5] the case $A = 1$, L arbitrary has been analyzed. A stability analysis shows that only stable or unstable nodes can be expected in this particular example. Nevertheless there arises a nontrivial migratory phase transition, with respect to the stability of a homogeneous population distribution, controlled by the agglomeration parameter $\varkappa^{\alpha\beta} = \varkappa$.

The case $A = 2$, $L = 2$ was analyzed in Weidlich and Haag[4] and partially in Chapter 4. It could be shown, that there exists a limit cycle solution under appropriate choice of the trend parameters.

In this chapter we study an example of a chaotic migratory flow pattern. This requires at least three independent variables in (6.11). For symmetrical reasons, we investigate the migration of three subpopulations, $A = 3$, between three regions, $L = 3$ within our dynamic framework of decision processes (Reiner et al[6]).

6.3 Chaotic Behaviour of Migratory Trajectories

We now specify the dynamic equations of motion (6.11) and study the particular case $A = 3$, $L = 3$ characterized by six independent variables $\overline{n}_{\alpha i}(t)$. Using the simplifying assumptions:

$$\delta_{\alpha i} = 0 \qquad \text{and} \qquad \nu_{ij}^{\alpha} = \nu \tag{6.14}$$

which means no preference for a certain region for any subpopulation and that the mobility for all individuals is the same (independent of distance), we obtain the specific set of dynamic equations of motion:

$$\frac{dx_{\alpha i}(t)}{d\tau} = \sum_{j=1}^{3} x_{\alpha j}(t)\exp[\sum_{\beta=1}^{3} \kappa^{\alpha\beta}(x_{\beta i}(t) - x_{\beta j}(t))]$$
$$- \sum_{j=1}^{3} x_{\alpha i}(t)\exp[\sum_{\beta=1}^{3} \kappa^{\beta\alpha}(x_{\beta j}(t) - x_{\beta i}(t))]$$
$$\equiv F_{\alpha i}(\boldsymbol{x})$$

for $\alpha = 1, 2, 3$, and $i = 1, 2, 3$, (6.15)

where we have introduced the scaled population numbers:

$$x_{\alpha i}(t) = \frac{\overline{n_{\alpha i}(t)}}{N_{\alpha}}, \quad \text{with} \quad 0 \le x_{\alpha i}(t) \le 1 \tag{6.16}$$

the scaled (dimensionless) time:

$$\tau = \nu\, t, \tag{6.17}$$

and scaled interaction parameters:

$$\kappa^{\alpha\beta} = N_{\alpha}\, \varkappa^{\alpha\beta}. \tag{6.18}$$

6.3.1 A Numerical Simulation

Using numerical solution methods we will investigate the dynamic behaviour of our system of interacting subpopulations. We assume for the inter-group and intra-group interaction parameters the following values:

$$(\kappa^{\alpha\beta}) = \begin{pmatrix} 1.7 & 1.5 & -1.5 \\ -1.5 & 1.7 & 1.5 \\ \kappa^{13} & -1.5 & 1.7 \end{pmatrix} \tag{6.19}$$

where the inter-group interaction parameter K^{13} will be changed from 1.5 to - 1.5, in order to demonstrate different possible solution patterns.

The results of the numerical simulations of (6.15) with the interaction matrice (6.19) is presented in a series of figures belonging to different values of K^{13}. The group of figures denoted by a) show the projection of the trajectories $x_{\alpha i}$, for α, i = 1, 2, 3, into the $\{x_{11}, x_{21}\}$ - plane. In the group b) of figures the logarithm of the *Fourier transform* $\hat{x}_{11}(f)$

$$\hat{x}_{11}(f) = \frac{1}{2\pi} \int_{-\infty}^{\infty} x_{11}(t)\, e^{-2\pi i f t} dt \tag{6.20}$$

of the variable $x_{11}(t)$ is depicted.

In Figs. 6.1 a,b the results are presented for K^{13} = 1.5. Obviously Fig.6. 1 a shows the projection of a simple limit cycle. Correspondingly, the Fourier spectrum, Fig.6.1 b, represents a few discrete frequencies belonging to this periodic migratory flow.

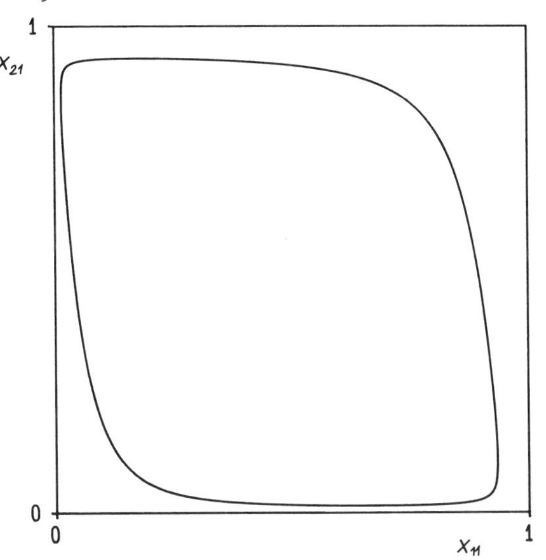

Figure 6.1 a: Projection of the stationary trajectory for K^{13} = 1.5

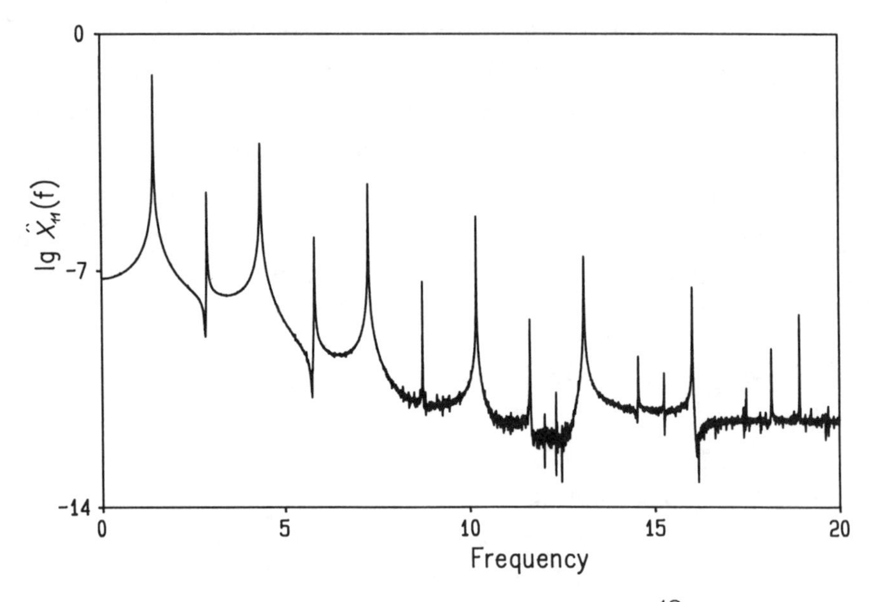

Figure 6.1 b: Fourier spectrum of the trajectory for $K^{13} = 1.5$

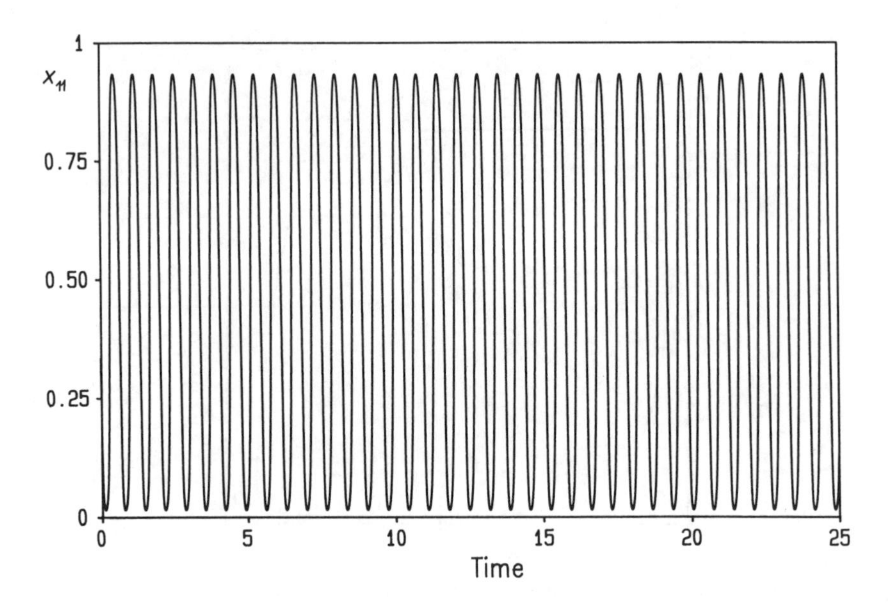

Figure 6.1 c: Evolution with time of $x_{11}(t)$ for $K^{13} = 1.5$

Figs. 6.2 a, b depict the behaviour of the trajectory for $K^{13} = -0.5$. Still there exists a limit cycle which, however, is more complex and contains two main periods.

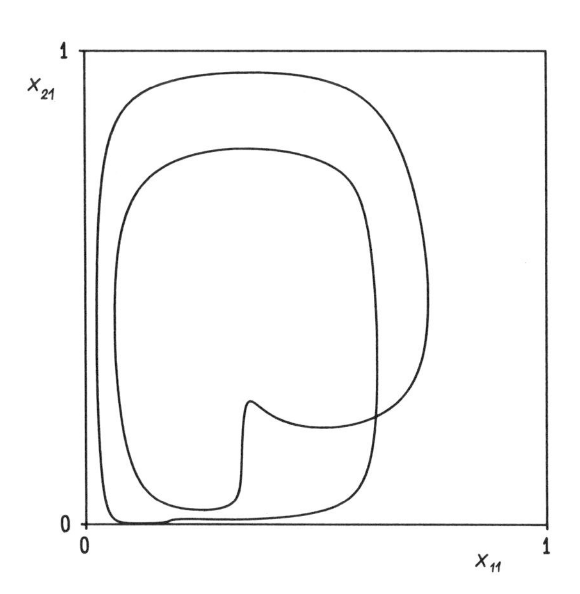

Figure 6.2 a: Projection of the stationary trajectory for $K^{13} = -0.5$

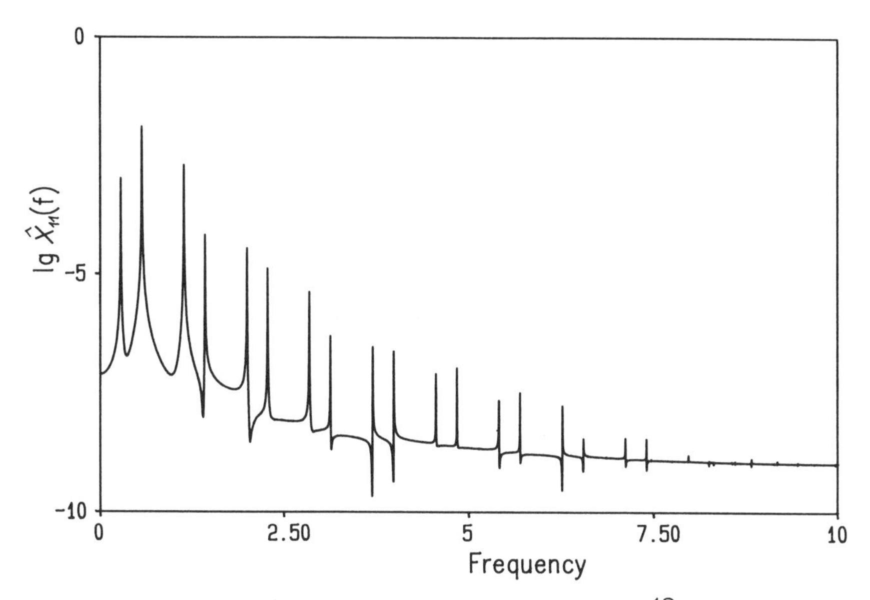

Figure 6.2 b: Fourier spectrum of the trajectory for $K^{13} = -0.5$

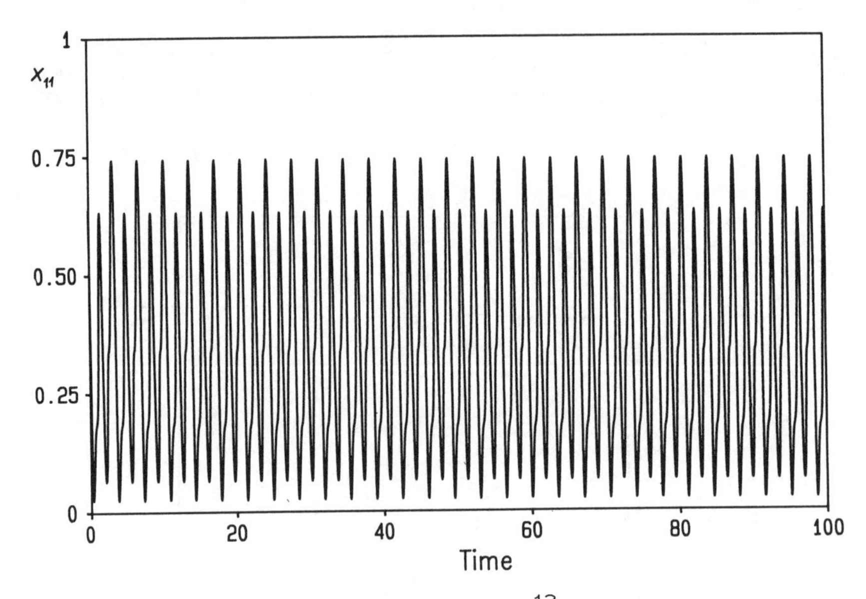

Figure 6.2 c: Evolution with time of $x_{11}(t)$ for K^{13}= - 0.5

Figs. 6.3 a, b show the solution pattern for the slightly modified value K^{13}= -0.55. There appears period doupling in the form of the limit cycle as well as in the Fourier spectrum.

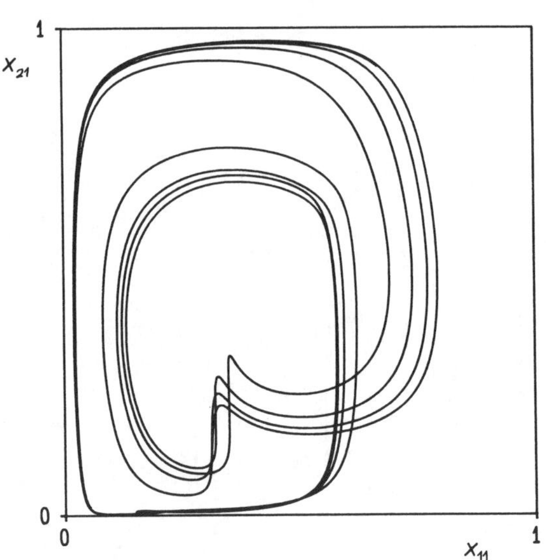

Figure 6.3 a: Projection of the stationary trajectory for K^{13}= - 0.55

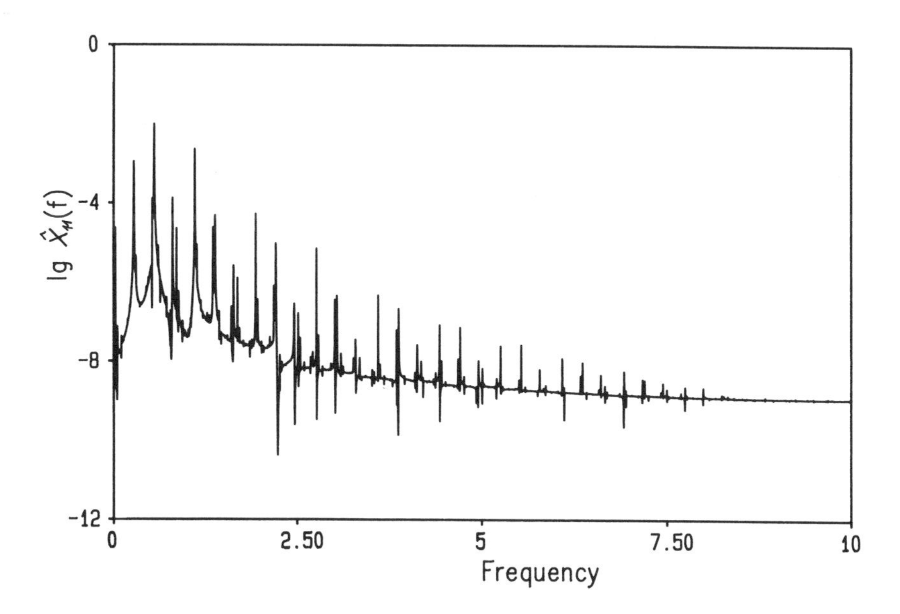

Figure 6.3 b: Fourier spectrum of the trajectory for $K^{13} = -0.55$

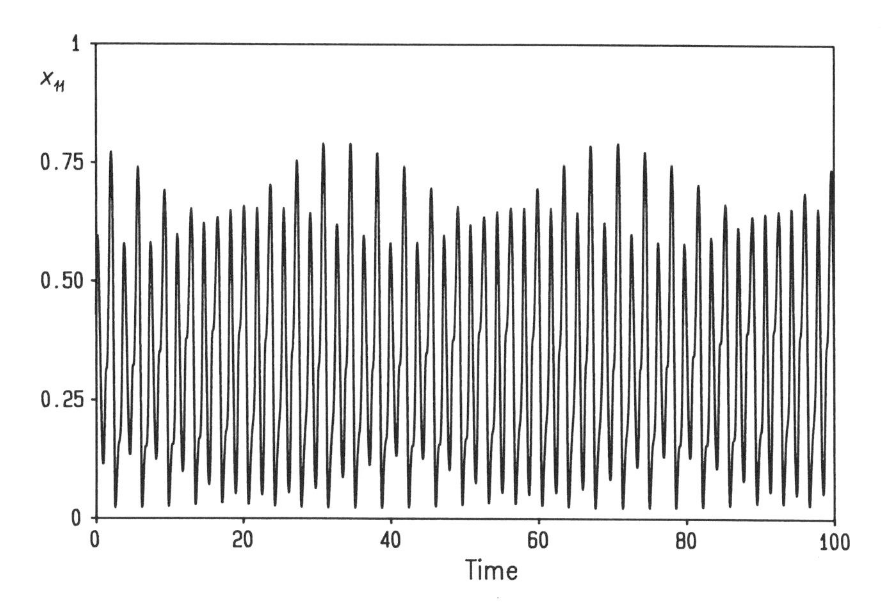

Figure 6.3 c: Evolution with time of $x_{11}(t)$ for $K^{13} = -0.55$

Finally, a chaotic dynamics is obtained for $K^{13} = -1.5$, as shown in Figures 6.4 a, b. The periodic limit cycle has now evolved into a nonperiodic strange attractor, characterized by a broad band continuous Fourier spectrum.

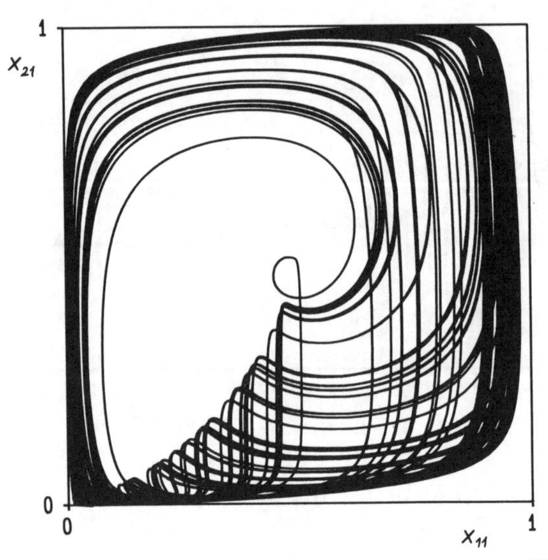

Figure 6.4a: Projection of the stationary trajectory for $K^{13} = -1.5$

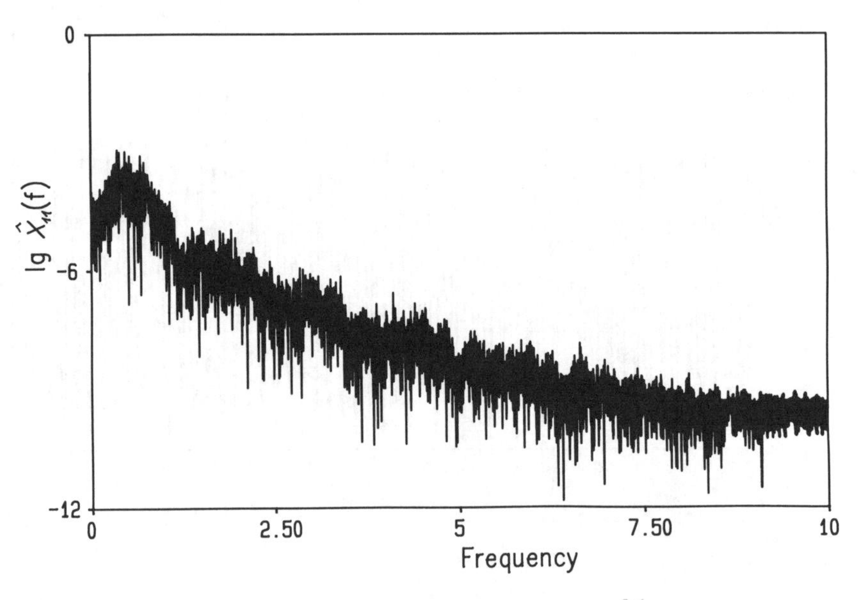

Figure 6.4b: Fourier spectrum of the trajectory for $K^{13} = -1.5$

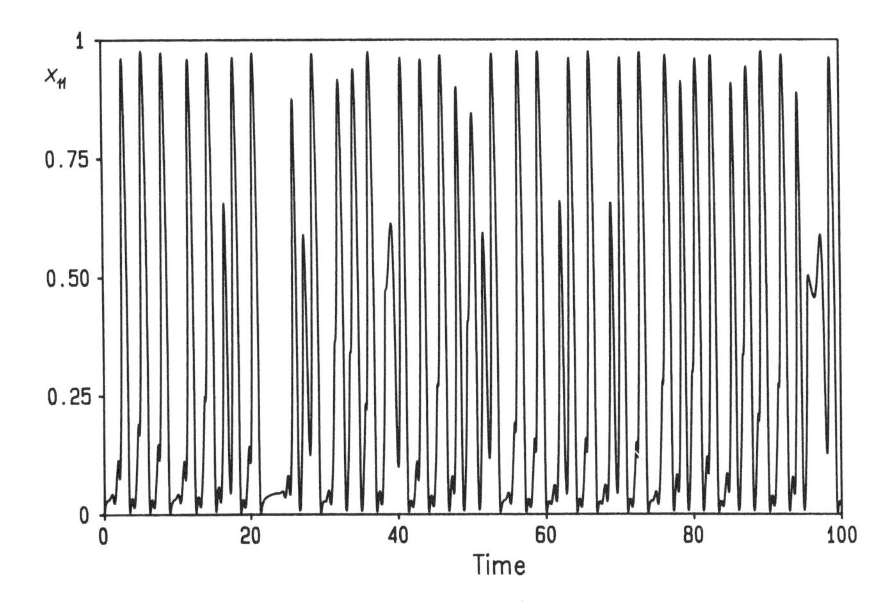

Figure 6.4c: Evolution with time of $x_{11}(t)$ for $K^{13} = -1.5$

The route to chaos pursued from Fig. 6.1 to Fig. 6.4 is called the *Ruelle-Takens picture*.

Let us now consider a further characterization of a strange attractor. For instance, we can ask whether or not a trajectory of a strange attractor fills out the whole state space or lies more or less on a hyperplane of lower dimension in state space. A first illustrative answer can be given in our example by considering Figures 6.5 a - d.

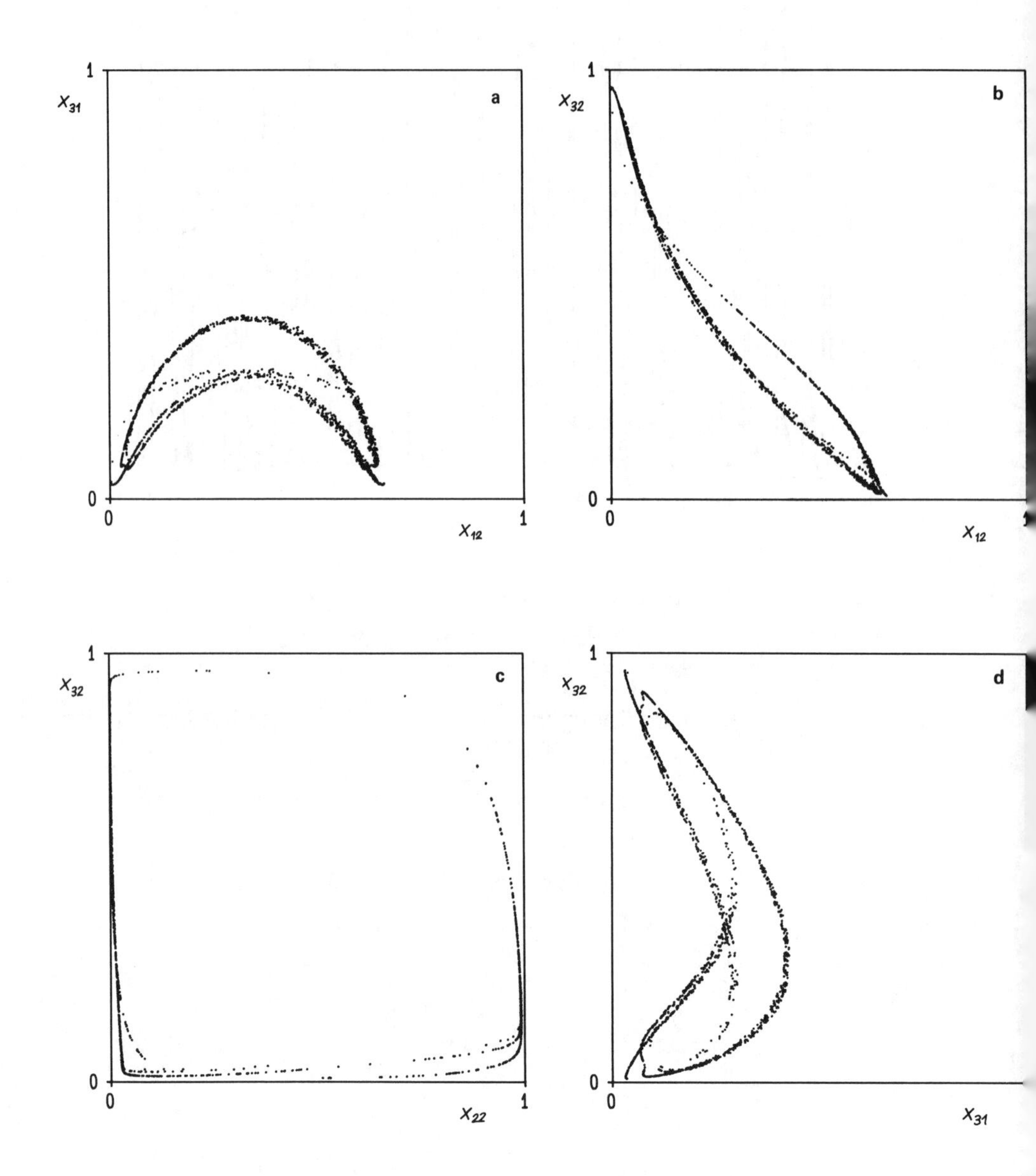

Figure 6.5: Projection of the traversing points of the state space trajectory of a strange attractor ($K^{13} = -1.5$) through the hyperplane $x_{11} = 1/3$ onto different planes

We shall register the values of all coordinates $\{x_{21}, x_{31}, x_{12}, x_{22}, x_{32}, x_{13}, x_{23}, x_{33}\}$, whenever the strange attractor crosses the hyperplane $x_{11} = 1/3$. Each crossing leads to one point in the $\{x_{12}/x_{31}\}$, $\{x_{12}/x_{32}\}$, $\{x_{22}/x_{32}\}$, $\{x_{31}/x_{32}\}$ projection of Figures 6.5 a-d, respectively. The points in the projections are distributed regularly, and each lies more or less on one-dimensional crossing lines. This means that the strange attractor itself more or less lies on an *almost* two-dimensional hyperplane instead of filling the whole 6-dimensional state space. Therefore, it seems appropriate to attribute an *effective dimension* to the strange attractor, which characterizes the subspace on which the motion of the trajectory takes place. Possible definitions of such a *fractal dimension* will be discussed in the next subsection.

6.3.2 Lyapunov Exponents and Fractal Dimensions

We shall now introduce a few mathematical concepts appropriate for the description of chaotic trajectories.

For certain domains of the state space the system (6.15) is *dissipative.* In these domains the flux in state space is *contractive.* That means, that a small volume element:

$$\Delta V = \prod_{\alpha=1}^{3} \prod_{i=1}^{3} \Delta x_{\alpha i} \tag{6.21}$$

contracts in the course of time. In other words

$$\frac{1}{\Delta V} \frac{d \Delta V}{d t} = \left[\prod_{\alpha=1}^{3} \prod_{i=1}^{3} \Delta x_{\alpha i}\right]^{-1} \frac{d}{d t} \left[\prod_{\alpha=1}^{3} \prod_{i=1}^{3} \Delta x_{\alpha i}\right] < 0 \tag{6.22}$$

Inserting of (6.15) in (6.22) yields that the contraction (6.22) is equivalent to:

$$\sum_{\alpha=1}^{3} \sum_{i=1}^{3} \frac{\partial F_{\alpha i}}{\partial x_{\alpha i}} < 0 \ . \tag{6.23}$$

Therefore, each volume element ΔV out of the dissipative domain is mapped to a zero volume for $t \rightarrow \infty$. The domain of volume zero, to which all trajectories starting within ΔV are finally attracted for $t \rightarrow \infty$ is denoted as *attractor*. Fixed points or limit cycles are specific examples of such attractors. If, however, the asymptotic dynamics are chaotic, we have the case of a so called strange attractor.

Chaotic solutions can be further analyzed by introducing the concepts of *Lyapunov exponents* (Haken[1]) and *fractal dimensions* (Hentschel and Procaccia[7]). These concepts were developed for the purpose of characterizing *strange attractors* in dynamic system theory.

Volume elements in dissipative systems are contractive (6.22), there need not be a contraction in *all* directions in phase space. Instead it is typical of a strange attractor that there exist stretching and folding processes for the volume element, and that the flux contracts in some directions and stretches in others.

In order to pursue this contraction process in detail let us consider an infinitesimal ball in (our 6-dimensional) state space.

During the dynamic evolution of our sytem (6.15), this volume element will be distorted, but, being infinitesimal, it will remain an ellipsoid. We denote the principal axes of this ellipsoid by $\varepsilon_i(t)$, i = 1, 2, ..., 6. It turns out that their evolution with time is given by:

$$\varepsilon_i(t) = \varepsilon_0(t) \exp(\lambda_i t) \tag{6.24}$$

From (6.24) we extract the definition of the *Lyapunov-exponents* λ_i

$$\lambda_i = \lim_{t\rightarrow\infty} \lim_{\varepsilon_i(0)\rightarrow 0} \left[\frac{1}{t} \ln\left(\frac{\varepsilon_i(t)}{\varepsilon_i(0)} \right) \right] \tag{6.25}$$

Because of the volume of the ellipsoid contracts for dissipative systems the relation:

$$\sum_{i=1}^{6} \lambda_i < 0 \tag{6.26}$$

must hold. Furthermore, the Lyapunov-exponents λ_i can be used to characterize the structure of the attractor:

1. If all Lyapunov-exponents are negative, $\lambda_i < 0$, for $i = 1, \ldots, 6$, the attractor is a stable focus
2. If $\lambda_1 = 0$, $\lambda_i < 0$, for $i = 2, \ldots, 6$, the attractor is a stable limit cycle. The principal axis of the ellipsoid along the limit cycle remains constant, while all other axes are contracted
3. If $\lambda_1 = \lambda_2 = 0$; and $\lambda_i < 0$, for $i = 3, \ldots, 6$, the attractor is a two-dimensional torus in phase space
4. If $\lambda_1 > 0$, a strange behaviour of the attractor is implied. Because of (6.22) at least one negative Lyapunov-exponent must exist. Hence, simultaneous stretching and contraction processes appear

Table 6.1

Lyapunov spectrum of system (6.15) for the interaction matrix (6.19)

K^{13}	λ_1	λ_2	λ_3	λ_4	λ_5	λ_6	D_{KY}
1.50	0.00	-3.56	-4.64	-6.15	-6.16	-8.89	1.00
0.00	0.00	-0.34	-0.34	-3.79	-8.30	-8.94	1.00
-0.50	0.00	-0.25	-0.25	-4.12	-7.56	-15.3	1.00
-0.55	0.00	-0.00	-0.79	-4.06	-7.10	-15.3	2.00
-0.60	0.48	0.00	-1.26	-4.11	-7.37	-14.1	2.38
-1.50	0.88	0.00	-7.54	-9.44	-11.9	-27.2	2.15

The Lyapunov-exponent for our migratory system has been computed using the algorithm described in Benettin et al.[8] The result is listed in Table 6.1.

It can be read off, that the characterization by Lyapunov-exponents comprizes the cases 2, 3 and 4 in full agreement with the trajectory analysis given in Figures 6.1 to 6.4. Furthermore, the *Kaplan-Yorke-dimension* (Kaplan and Yorke[9]) D_{KY} of an attractor is also listed in Table 6.1.

The Kaplan-Yorke-dimension is defined as a function of the Lyapunov-exponent in the following way:

$$D_{KY} = j + \frac{\sum_{i=1}^{j} \lambda_i}{|\lambda_{j+1}|} \tag{6.27}$$

Here, the exponents are ordered in descending order $\lambda_1 \geq \lambda_2 \geq \ldots \geq \lambda_6$, and j is the largest integer, for which $\sum_i^j \lambda_i \geq 0$. Of course, the dimension D_{KY} is an integer for a limit cycle ($D_{KY} = 1$) or for a torus in phase space ($D_{KY} = 2$), but it can be fractal for a strange attractor.

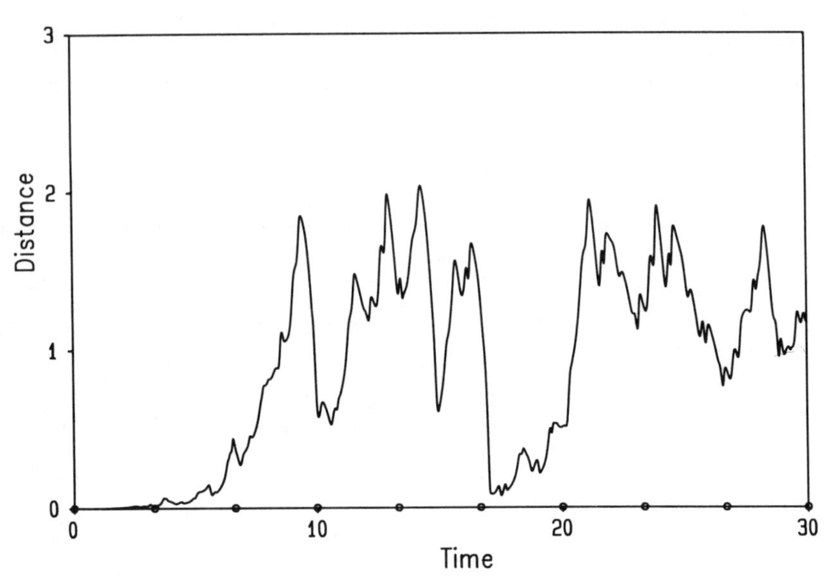

Figure 6.6 : Distance of two initially adjacent trajectories in the case of a limit cycle ($K^{13} = 1.5$, —o—o—) and of a strange attractor ($K^{13} = -1.5$, ———)

As a consequence of the stretching processes in the case of strange attractors initially infinitesimally neighbouring trajectories separate exponentially in time. This is demonstrated in Fig. 6.6, where the euclidian distance in phase space of two trajectories has been plotted in the case of a limit cycle ($K^{13} = -1.5$) and for the strange attractor ($K^{13} = 1.5$).

Finally we consider another measure of the effective dimension of an attractor, namely the *correlation dimension*, which has been introduced by Grassberger and Procaccia.[10] This intuitively appealing dimension is obtained from the correlation between random points on the attractor.

Consider a set $\{\mathbf{y}_i\,,\ i = 1, 2, \ldots, M\}$ of points on the attractor, obtained e.g., from a time series $\mathbf{y}_i = \{x_{\alpha k}(t + i\tau)\}$ with a fixed time increment τ between successive measurements. Due to the exponential divergence of trajectories, the pairs of points will be dynamically uncorrelated, but they are spatially correlated, since they are points on the attractor. This spatial correlation can be measured by a sum $C(r)$, defined as:

$$C(r) = \lim_{M\to\infty} \frac{1}{M^2} \left[\begin{array}{l} \text{number of pairs } (i,j) \text{ with} \\ \text{euclidian distance } |x_i - x_j| < r \end{array} \right]. \tag{6.28}$$

or equivalently:

$$C(r) = \lim_{M\to\infty} \frac{1}{2M^2} \left[\sum_{i,j=1}^{M} \Theta(r - |\mathbf{x}_i - \mathbf{x}_j|) - M \right]. \tag{6.29}$$

If the attractor densely fills a subspace of the phase space of dimension D_c it is clear that $C(r)$ should grow with r as follows:

$$C(r) \sim r^{D_c}, \tag{6.30}$$

since also the random points j on the attractor contained in the ball of radius r around the reference point i fill that ball on a subspace of dimension D_c. Therefore, their number should grow accordingly (6.30).

Calculating $C(r)$ for sufficiently large numbers M and plotting $\ln C(r)$ versus $\ln r$, the correlation dimension D_c is obtained as the slope of the curves. This has been done in Fig. 6.7 for the case of a limit cycle ($K^{13}=1.5$), where $D_c = 1$, and for the case of a strange attractor ($K^{13} = -1.5$), where we obtain $D_c \approx 2.13$. The latter result is in best agreement with the observation made in Fig. 5, that the strange attractor lies almost on a two-dimensional surface within the phase space. A variety of other definitions of fractal dimensions has been described in Farmer et al.[11]

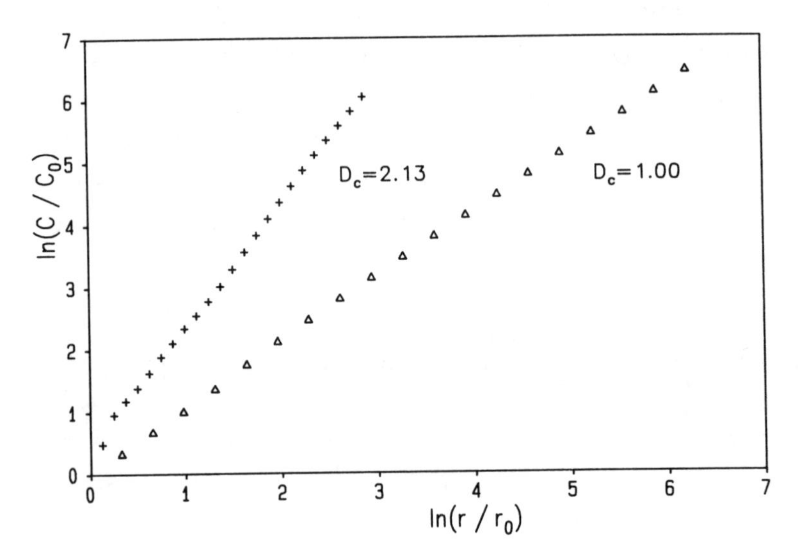

Figure 6.7: Determination of the correlation dimension D_c of a limit cycle ($K^{13} = 1.5$, Δ) and of a strange attractor ($K^{13} = -1.5$,+)

Grassberger and Procaccia have proved that the relation:

$$D_{KY} \geq D_c \tag{6.31}$$

must hold between the two *fractal dimensions*. This relation is also confirmed by the numerical results.

However, sometimes one has no access to a time series of *all* coordinates (variables of the system) of the attractor in the phase space. Even under such restrictive conditions it is possible to determine D_c using a data set consisting of only *one* variable. In that particular case one constructs *artificial* d-dimensional vectors from the available time series of data by:

$$\mathbf{y}_i = \{x(t_i),\ x(t_i+\tau),\ \ldots,\ x(t_i+(d-1)\tau)\}\,. \tag{6.32}$$

The correlation sum $C(r)$ can then be calculated according to (6.28) or (6.29) , using the d-component vectors $\mathbf{y}_i$ instead of the *real* $\mathbf{x}(t)$. If d is equal or larger than the true dimension of the system, to whose attractor the component $\mathbf{x}(t)$ belongs, the same D_c should be obtained. This statement is checked in Figures 6.8, 6.9.

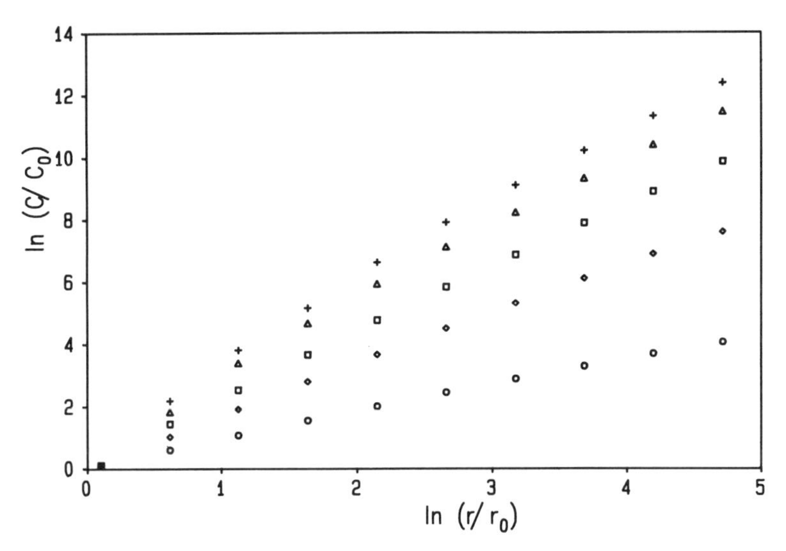

Figure 6.8: Correlation integral $C(r)$ versus r for different values d. (d = 1, o), (d = 2, ◇), (d = 3, □), (d = 4, △), (d = 5, +)

Fig. 6.8 shows the result of the evaluation of $C(r)$ using the time series $x_{11}(t_i)$ with d = 1, 2, ..., 5. In Fig. 6.9 the correlation dimension D_c obtained from Fig. 6.8 is plotted versus the dimension d of the space into which the system is embedded.

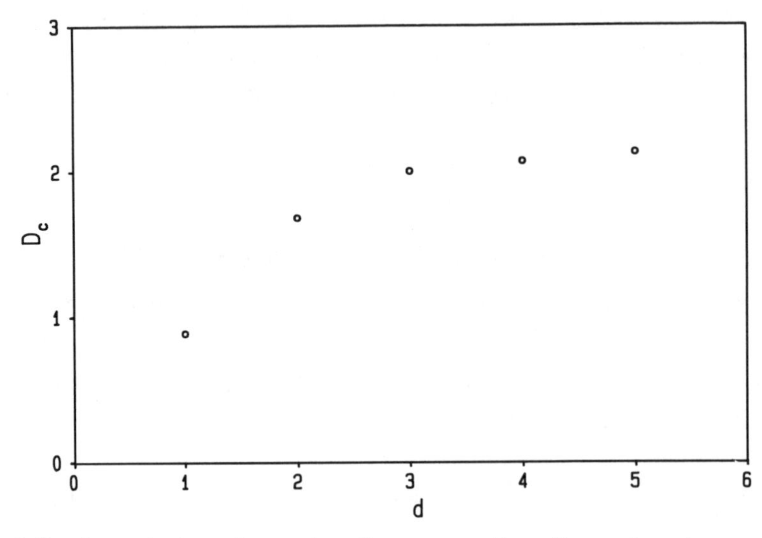

Figure 6.9: Correlation dimension D_c versus the dimension d

6.4 Conclusion

The investigation in this chapter has shown, that migratory systems with nonlinear transition rates provide another example to which the concepts of strange attractors and deterministic chaos fully apply under certain trend parameter conditions. Under such circumstances migratory systems, even if treated with quasi-deterministic equations, may behave highly irregular and unpredictable. From the results presented in Chapter 6 and further investigations we can formulate necessary conditions for the occurrence of chaotic trajectories:

1. The intra-group interaction must be positive and above a critical value
2. The sign of the inter-group interactions between at least two subpopulations must be different
3. The interaction matrix must be unsymmetric, since for an antisymmetric interaction matrix only stable fixpoints or limit cycles exist

This demonstrates that an endogeneous migratory system exhibits chaotic solutions under very restrictive conditions only. In most of real world migratory systems. however, we expect a regular behaviour of the migration flows. Nevertheless, chaos can be induced even under such conditions by exogeneous changes of the socio-economic situation of a country.

Chapter 7

Spatial Interaction Models and their Micro-Foundation

Implicit in all spatial urban sub-models is the need to allocate and match *demand* and *supply* ; *needs* to *facilities*. In the last decades much work focused on different approaches to this issue . Recently the main broad approaches have been reviewed in order to assemble larger components to an *integrated urban model* (for details see Bertuglia et al[1]). Two main building blocks underpin this approach from a macroeconomic point of view:

1. The definition and identification of the major *markets* or allocation systems; e.g., the *housing* and *labour* market, *land*, *services* and *transport*

2. The selection of adequate (empirically accessible or at least computable) major socio-economic *macrovariables*, e.g., *population*, *stocks* or *facilities*, and *prices*

On the other hand, the basic elements of microeconomic choice theory deal with a quite different approach:

1. The definition of *individual agents* e.g., households , landlords , workers , consumers
2. The selection and construction of an adequate *choice strategy* with respect to certain constraints , e.g., profit maximization, utility optimization

The achievement of planning goals when applying land-use and pricing policy instruments in the strategic planning of user-attracting spatial systems, such as retail centres or hospitals, is inevitably coupled with the preferences of users in their choice of the competing facilities in space (see Roy[2,3]). In Leonardi,[4,5] an important distinction was made between two classes of urban service systems: *delivery systems* where the same decision-maker is responsible for both locating a service facility and allocating demand to it, and *user-attracting systems*, where different decision-makers locate the facilities compared with those demanding the facility services. Delivery systems, such as emergency health care services, are traditionally designed by techniques out of the field of *operations research*. The design of user-attracting systems, such as networks of hospitals or retail centres, require a more sophisticated treatment, because of the presence of multiple decision-makers in the system. Methods which account for conflicting objectives of different decision-makers e.g., customers, retailers, patients, health authorities ect., have recently been developed. Mayhew and Leonardi[6] worked on embedding concepts, Roy and Lesse[7] used game theory to overcome these difficulties, and in Roy et al[8] methods from oligopoly analysis have been taken.

It is the aim of this chapter to concentrate on this more challenging problem, namely the design of user-attracting spatial systems. It will be demonstrated that this dynamic choice theory can be used as a good starting point for the construction of an *integrated urban model.*

The (micro- and macro-) economic approach establishes in some sense the foundations for other approaches by discussing mechanisms for adjustment of both quantities and prices. The earliest consideration of a user-attracting system balancing customer and retailer behaviour appears to be Lakshmanan and Hansen's [9] equilibrium model. In the fundamental work of Wilson,[10] Harris and Wilson[11] and Allen and Sanglier[12] a particular set of spatial-interaction based adjustment methods for stocks has been established. Particular emphasis is given in these models to supplyside structures and to the way in which these pattern evolve and change.

The integrated system has to be knitted together by a suitable set of probabilistic accounts. The broadest framework for this is offered by the master equation approach. With reasonable additional assumptions this dynamic choice theory can be applied in this way.

Contrary to the two briefly described procedures below we start from the microeconomic decision process of individual agents and add socio-economic mechanisms to our probabilistic picture. Thus the link between the *microlevel* and the *macrolevel* can be established.

7.1 Introduction to Spatial Urban Theory

The early urban models of Lowry,[13] Hill,[14] and Forrester[15] have often been criticized and essentially ignored by economists. The main point of criticism stressed that they were not based on economic theory (see Mills[16]). Therefore, practical questions such as doing cost-benefit analyses or predicting the response of real estate prices to public investment could not be answered. On the other hand, these models initiated a great amount of creative work in this field.

Allen and Sanglier introduced a spatial interaction model where jobs and populations of different kinds follow a logistic evolution path in each region and compete with each other via corresponding *capacities*. Concepts like *capacities*, distance-effects, and some economic aspects have been included in the model. Of course, such a logistic model structure can also be derived from a stochastic background theory assuming a birth/death-structure in the transition rates. However, not all processes which are important in spatial dynamics can be fitted into this restrictive frame, since migration or flow-processes are important as well.

Harris and Wilson's spatial interaction model is also of a logistic type. The imbalance between revenue attracted to a region and the cost of supply to maintain a certain level of the facility stock leads to changes in the facility stock. Contrary to the above model, however, Harris and Wilson take into account expenditure flows, but neglect on the other hand some important economic facts which are included in the first model. The expenditure flows are directly linked to the facility stock taking into account an interaction decreasing with distance. It was demonstrated, that the achievement of a customer/retailer equilibrium is consistent with the maximization of a certain economic/consumer surplus, entropy or group accessibility measure for the customers in the system.

These two important papers raised the interest of geographers and economists to apply these models to real world situations and to compare the empirical trajectories with the different model outputs. However, three main problems became obvious:

1. The first problem is caused by the nonlinear structure of both models (see Pumain,[17] Pumain, Saint-Julien and Sanders,[18] Lombardo and Rabino[19]). Considerable difficulties in calibration exist, since the number of trajectories which can be compared with empirical data is smaller than the number of fitting parameters. Therefore, in a simple application of the models there is no general procedure to guarantee a unique choice of the trend parameters which have to be estimated. Unfortunately an uncertainty in the model parameters may lead to unacceptable differences in the trajectories even in a short-time simulation
2. Both models mainly focus on the supply-side. There are no dynamic equations for the demand-side, which could describe dynamic reactions of the expenditure flow to changes in the facility stock
3. The dynamic equations of motion of the models discussed so far are not derived from microeconomic foundations starting from the decision behaviour of the different decision-makers

In this chapter we improve the construction of spatial-interaction models using our framework of a dynamic decision theory. We are mainly following the argumentation in Haag,[20] and Haag and Wilson. [21] Our starting points are the decision processes of the individuals who decide to use facilities of different regions as well as of the entrepreneurs, retailers and landowners who decide about changes in the facility stock, the prices of goods and services and the land rents. We shall not use any equilibrium assumption in deriving the fundamental dynamic equations of motions for the relevant macro-variables of the system. This broader overall framework with full probabilistic dynamic underpinning also facilitates extensions in a number of directions. For instance, it offers a simple way of introducing saturation effects. It also becomes obvious how conventional spatial-interaction models can be generalized, especially how fundamental dynamic aspects can be introduced. The master equation offers a new approach to parameter estimation and the known bifurcation properties of spatial-interaction models can be used to see how these properties manifest themselves in master equations.

7.2 A Service System as the Basis of the Model

We choose the simplest example in order to illustrate the main principle to be explored in this chapter. However, the approach can be extended in a straightforward manner.

We consider four kinds of *agents* (see Table 7.1) whose decisions control destination choice (*consumers*), facility size (*entrepreneurs*), prices of goods and services (*retailers*) and land rent (*landowners*).

Table 7.1

The different agents of the service sector model

agent	decisions about	controlled variable	
consumer	destination choice	expenditure flow	$T_{ij}(t)$
developers	facility size	facility stock	$Z_j(t)$
retailers	prices of goods and services	price level	$p_j(t)$
land owners	rent of land	rent level	$r_j(t)$

We consider an urban system consisting of L zones with a given transportation network. The transportation costs between zone i and zone j are denoted by c_{ij}. The *expenditure configuration* is described by the array $\underline{T}(t) = \{T_{ij}(t)\}$ at time t. We will assume that this array is determined by utility functions $v_{ij}(t)$ through a mechanism to be described later. The element $T_{ij}(t)$ describes the expenditure flow from a zone i (the consumers' residence) to zone j (where the consumer is using facilities) at time t. Moreover the utility functions $v_{ij}(t)$ have two subscripts since they relate to both housing and service sectors.

The total revenue $D_j(t)$ attracted to each zone j is obtained by:

$$D_j(t) = \sum_{i=1}^{L} T_{ij}(t) \tag{7.1}$$

and the total expenditure $E_i(t)$ of consumers living in zone i is:

$$E_i(t) = \sum_{j=1}^{L} T_{ij}(t) \tag{7.2}$$

In general, a number of variables like $v_{ij}(t)$ may be functions of arrays such as $\underline{T}$, involving a high degree of feedback, but we only show such interdependence here when it is required in this simple example. But it may be necessary to add it again if the model is extended.

The utility functions $v_{ij}(t)$ are assumed to depend on the supply side variables. Let $Z_j(t)$ be the scale of provision of facilities in zone j at time t and let $p_j(t)$ be the price for one unit of the service (assuming one type of good or service for simplicity). Let $r_j(t)$ be the land rent to be paid per unit of facilities provided at j.

The next step to be solved is a subsidiary problem: the dynamics of the price adjustment process. If a fast dynamic adaptation of the monetary variables (prices and rents) is assumed, they can be replaced by "relative" *equilibrium* values (this means equilibrium with respect to other socio-economic variables, and thus being a moving equilibrium). Then one possible answer to the question of how consumers at i perceive the prices of goods at j is something like $p_j + c_{ij}$, where c_{ij} is the *unit* transport cost connected with buying a unit service (or good) in region j.

In general, however, this "adiabatic" elimination of the fast variables, namely the prices and land rents cannot be justified, e.g., during a fast expansionary phase of the whole spatial-temporal system. In such a case a complicated mutual interaction between all socio- economic variables determines the dynamics.

Therefore the full modelling problem becomes: how can the destination choice at time t, represented as $\{T_{ij}(t)\}$, the spatial and temporal pattern of provision $\{Z_j(t)\}$, the prices $\{p_j(t)\}$ and land rents, $\{r_j(t)\}$ be determined for given initial conditions $\{T_{ij}(0)\}$, $\{Z_j(0)\}$, $\{p_j(0)\}$ and $\{r_j(0)\}$, respectively?

In the next section, a master equation approach to this problem will be presented.

7.3 A Master Equation Approach

Decision processes are stochastic processes, since we are interested in the probability $P(\boldsymbol{c}, t)$ that a configuration $\boldsymbol{c} = \{\underline{T}, \boldsymbol{Z}, \boldsymbol{p}, \boldsymbol{r}\}$ is realized at time t. We have seen (Chapter 2) that the master equation is the equation of motion for this probability distribution:

$$\frac{dP(\boldsymbol{c},t)}{dt} = \sum_{\boldsymbol{k}} w(\boldsymbol{c},\boldsymbol{c}+\boldsymbol{k})P(\boldsymbol{c}+\boldsymbol{k},t) - \sum_{\boldsymbol{k}} w(\boldsymbol{c}+\boldsymbol{k},\boldsymbol{c})P(\boldsymbol{c},t) \tag{7.3}$$

with:

$$\sum_{\boldsymbol{c}} P(\boldsymbol{c},t) = 1 \ . \tag{7.4}$$

In the configuration space the maximum (or the maxima) of $P(\boldsymbol{c}, t)$ represents the most probable expenditure flow pattern and the corresponding spatial distribution of the facility stock, the prices and rents. In Fig. 7.1 schematically one possible state $\boldsymbol{c}$ is represented which can be found with probability $P(\boldsymbol{c}, t)$.

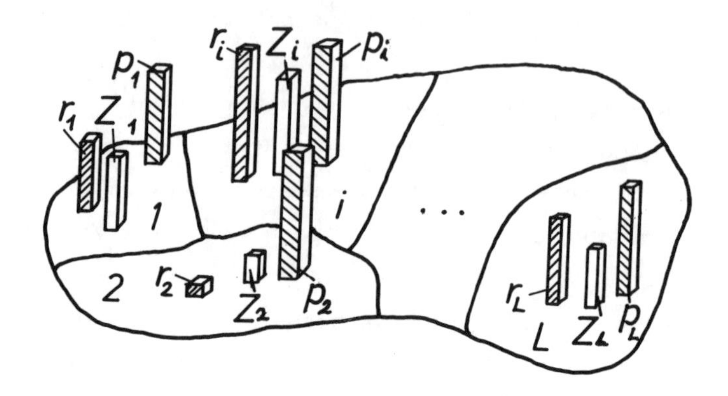

Figure 7.1: A possible spatial distribution of the facility stock $\boldsymbol{Z}$, the rents $\boldsymbol{r}$, and price-level $\boldsymbol{p}$ in a system of L regions

Since the master equation (7.3) contains the full stochastic information on the system, not only the most probable decision (or socio) configuration or its mean values can be obtained but also the variances of the distribution function.

The distribution function $P(\boldsymbol{c}, t)$ does not factorize in general,

$$P(\boldsymbol{c}, t) \neq P^{(1)}(\underline{T}, t) P^{(2)}(\boldsymbol{Z}, t) P^{(3)}(\boldsymbol{p}, t) P^{(4)}(\boldsymbol{r}, t) \tag{7.5}$$

because of the transition rates $w(\boldsymbol{c} + \boldsymbol{k}, \boldsymbol{c})$ are simultaneously dependent on *all* variables $\boldsymbol{c}$. It can be proved that the condition of detailed balance is not fulfilled for this model on the service sector.

7.3.1 The Total Transition Rates for the Service Sector Model

The total transition rate $w(\boldsymbol{c} + \boldsymbol{k}, \boldsymbol{c})$ is obtained as the sum over contributions of different socio-economic processes:

$$w(\mathbf{c}+\mathbf{k},\ \mathbf{c}) = \sum_{i,j,k}^{L} w^{(1)}_{ik,ij}(\underline{T}+\underline{k}^{(1)}, \mathbf{Z}, \mathbf{p}, \mathbf{r}; \underline{T}, \mathbf{Z}, \mathbf{p}, \mathbf{r})$$

$$+ \sum_{j}^{L}[w^{(2)}_{j+}(\underline{T}, \mathbf{Z}+\mathbf{k}^{(2)}, \mathbf{p}, \mathbf{r}; \underline{T}, \mathbf{Z}, \mathbf{p}, \mathbf{r}) + w^{(2)}_{j-}(\underline{T}, \mathbf{Z}+\mathbf{k}^{(2)}, \mathbf{p}, \mathbf{r}; \underline{T}, \mathbf{Z}, \mathbf{p}, \mathbf{r})]$$

$$+ \sum_{j}^{L}[w^{(3)}_{j+}(\underline{T}, \mathbf{Z}, \mathbf{p}+\mathbf{k}^{(3)}, \mathbf{r}; \underline{T}, \mathbf{Z}, \mathbf{p}, \mathbf{r}) + w^{(3)}_{j-}(\underline{T}, \mathbf{Z}, \mathbf{p}+\mathbf{k}^{(3)}, \mathbf{r}; \underline{T}, \mathbf{Z}, \mathbf{p}, \mathbf{r})]$$

$$+ \sum_{j}^{L}[w^{(4)}_{j+}(\underline{T}, \mathbf{Z}, \mathbf{p}, \mathbf{r}+\mathbf{k}^{(4)}; \underline{T}, \mathbf{Z}, \mathbf{p}, \mathbf{r}) + w^{(4)}_{j-}(\underline{T}, \mathbf{Z}, \mathbf{p}, \mathbf{r}+\mathbf{k}^{(4)}; \underline{T}, \mathbf{Z}, \mathbf{p}, \mathbf{r})] \tag{7.6}$$

The terms $w^{(1)}_{ik,ij}$ refer to changes in the expenditure flow configuration $\underline{T}$, due to decisions of consumers to change from a state having residence in i and using facilities in j to a state of still having residence in i but now using facilities in k. Changes of residential location are not taken into account, since it is reasonable to assume that the housing mobility will be considerably slower than the shopping mobility of the population. Indeed decisions to buy in another zone can be made much easier than moving from one flat to another (see Leonardi [22]). This means it is assumed that the spatial distribution of households can be given exogenously in relation to determining the location of facilities. It is well known that households make their location decisions not only with respect to their places of work, but also in relation to public and private facilities. Thus, the problem cannot be so readily decoupled. Feed-back effects should be allowed between private residential location decisions and the location decisions for private and public facilities. However, the assumption of a fixed residential location can easily be relaxed (compare Haag[20]).

The $w^{(l)}_{j+}$, $w^{(l)}_{j-}$ describe decisions of entrepreneurs ($l = 2$), retailers ($l = 3$) and landowners ($l = 4$) to make available or to remove one unit of the facility stock Z_j , and to increase or decrease the prices p_j and the rents r_j , respectively.

In (7.6) we have neglected contributions of simultaneous transitions to the total transition. This means that a transition to a neighbouring state in the configuration space is assumed as a sequential process of single steps of changes in consumers' flows, facility stocks, prices and rents, instead of a simultaneous transition of all of them. Later we introduce scaled variables. The transition rates then are functions of these scaled variables. The single step transitions from a decision configuration $\boldsymbol{c}$ to a neighbouring one $\boldsymbol{c} + \boldsymbol{k}$ are depicted in Figure 7.2.

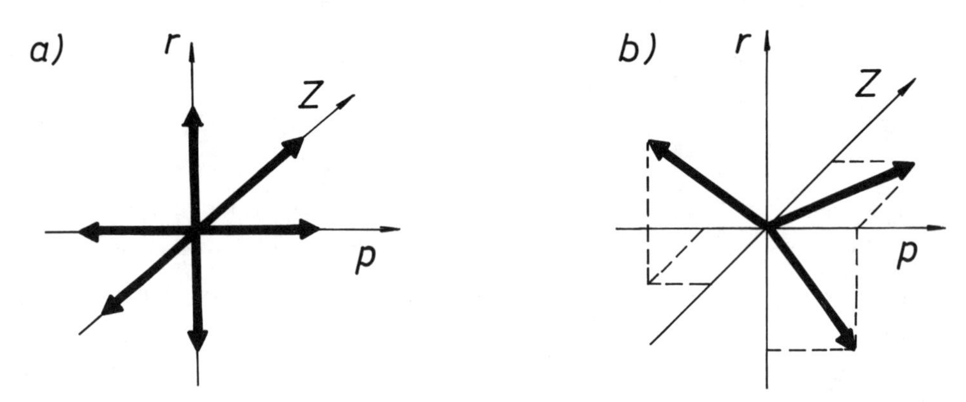

Figure 7.2: (a) Transitions of the aggregate system to neighbouring states which are taken into account
(b) A few simultaneous transitions which are neglected

The next step is to introduce specifications of the individual transition rates. Of course, the decisions of each group of agents will depend on the spatial decision pattern of the other groups.

7.3.2 Consumer Dynamics

Let $n_{ij}(t)$ be the number of consumers having residence in i and using facilities in j. Then:

$$n_i(t) = \sum_{j=1}^{L} n_{ij}(t) \tag{7.7}$$

is the number of consumers living in i, and the total number of consumers is given by:

$$N(t) = \sum_{i=1}^{L} n_i(t). \tag{7.8}$$

The individual transition rate $p_{ik,ij}(\boldsymbol{c})$ for a consumer living in i for changing his shopping area $j \rightarrow k$ is assumed to depend on the expected utility gain $(v_{ik} - v_{ij})$:

$$p_{ik,ij}(\boldsymbol{c}) = \varepsilon_1(t) \exp[v_{ik}(t) - v_{ij}(t)] \tag{7.9}$$

where $\varepsilon_1(t)$ is a time scaling parameter. Then the total number of changes of consumer trips $(i, j) \rightarrow (i, k)$ is given by:

$$\tilde{w}^{(1)}_{ik,ij} = n_{ij}(t)\, p_{ik,ij}(\boldsymbol{c}). \tag{7.10}$$

In many applications it is reasonable to assume that the expenditure flows $T_{ij}(t)$ can be linked to the consumer numbers $n_{ij}(t)$ via:

$$T_{ij}(t) = g(t) n_{ij}(t), \tag{7.11}$$

where $g(t)$ can be seen as the average of individual needs. By a comparison of the total stocks $g(t)$ can be easily determined. Thus we are able to derive from the consumer dynamics (7.10) transition rates per unit of time for the expenditure flows:

$$w^{(1)}_{ik,ij}(\underline{T}+\underline{k}^{(1)}, \boldsymbol{Z}, \boldsymbol{p}, \boldsymbol{r}; \underline{T}, \boldsymbol{Z}, \boldsymbol{p}, \boldsymbol{r}) = \begin{cases} T_{ij}\, \varepsilon_1(t) \exp[v_{ik} - v_{ij}] \\ \quad \text{for } \underline{k}^{(1)} = \{0,..,1_{ik},..,(-1)_{ij},..,0\} \\ 0 \quad \text{for all other } \underline{k}^{(1)} \end{cases} \tag{7.12}$$

Expression (7.12) can be inserted into the master equation (7.3), this gives the evolution with time to find a certain expenditure flow pattern. However, the facility stock, the prices and rents interact with these flow patterns. Therefore, it is essential to consider the decision processes of the other agents too.

7.3.3 Decision Processes of Developers, Retailers and Land Owners

The dynamics of the provision of facilities $\boldsymbol{Z}$, the prices $\boldsymbol{p}$, and rents $\boldsymbol{r}$ can be considered together because the same kind of methods can be used. In particular, because there is no *migration* or *flow* involved, they are each considered formally as birth/death processes, that means of elementary steps of increase or decrease, respectively, of the corresponding quantities.

Let $\boldsymbol{X}^{(l)} = \{X_1^{(l)}, X_2^{(l)}, \ldots, X_L^{(l)}\}$ be the facility stock configuration ($l = 2$, $X_j^{(2)} = Z_j$), the price configuration ($l = 3$, $X_j^{(3)} = p_j$), or the rent configuration ($l = 4$, $X_j^{(4)} = r_j$), respectively. By decisions of the corresponding agents, the configuration $\boldsymbol{X}^{(l)}$ will change in the course of time. As above, we will again introduce individual transition rates to these processes. Let $w_{j+}^{(l)}(\boldsymbol{X}^{(l)} + \boldsymbol{k}^{(l)}, \boldsymbol{X}^{(l)})$, $w_{j-}^{(l)}(\boldsymbol{X}^{(l)} + \boldsymbol{k}^{(l)}, \boldsymbol{X}^{(l)})$ be the *birth rate* and *death rate* per unit of time of the stock variable $\boldsymbol{X}^{(l)}$. Then a rather general formulation for these rates can be seen in the following ansatz (cf., Haag,[20] Haag and Wilson[21]):

$$w_{j\pm}^{(l)}(\boldsymbol{X}^{(l)}+\boldsymbol{k}^{(l)}, \boldsymbol{X}^{(l)}) = \begin{cases} 1/2\, f_j^{(l)}(\boldsymbol{X}) \exp[\pm \Phi_j^{(l)}(\boldsymbol{c})] & \text{for } \boldsymbol{k}^{(l)} = \{0, \ldots, (\pm 1)_j, \ldots, 0\} \\ 0 & \text{for all other } \boldsymbol{k}^{(l)} \end{cases} \tag{7.13}$$

for $l = 2, 3, 4$ and for rather general functions $f_j^{(l)}(\boldsymbol{X}) > 0$ and $\Phi_j^{(l)}$. Of course, in the transition rates (7.13) an " immigration" term could be included which describes the possible settlement of facilities in - initially - empty zones. The functions can be modelled to include saturation effects of the various parameters.

The factor $f_j^{(l)}(\boldsymbol{X})$ describes the speed of adjustment. Since the birth/

death rates must not be negative, the condition $f_j^{(l)}(\mathbf{X})>0$ has to be fulfilled. Here we assume for simplicity that the speed of adjustment depends on the already realized scale of the stock $\mathbf{X}^{(l)}$ in a linear way, with time scaling parameters $\varepsilon_l(t)$:

$$f_j^{(l)}(\mathbf{X}) = \varepsilon_l(t) X_j^{(l)} . \tag{7.14}$$

The function $\Phi_j^{(l)}(\mathbf{c})$ takes into account the imbalance between cost of supply and the revenue attracted to facilities in the zone under consideration. If there is an economic surplus $\Phi_j^{(l)}(\mathbf{c}) > 0$, (or deficit $\Phi_j^{(l)}(\mathbf{c}) < 0$) it is likely that the facility stock is expanded, the prices and the land rents are increased (or vice versa, that the facility stock is reduced, the relative prices and land rents are lowered).

Reasonable assumptions are:

$$\Phi_j^{(l)}(\mathbf{c}) = \lambda_l(t)[D_j(t) - C_j(t)] \tag{7.15}$$

or alternatively:

$$\Phi_j^{(l)}(\mathbf{c}) = \lambda_l(t)[D_j(t) - C_j(t)]/Z_j(t) \tag{7.16}$$

Here, $\lambda_l(t) > 0$ describes the intensity of response of an *agent* to an economic surplus $\Phi_j^{(l)} > 0$. The total revenue attracted to j is $D_j(t)$, and the total cost of supplying facilities of size $Z_j(t)$ at zone j, where the unit land rent is $r_j(t)$, is denoted by $C_j(t)$. We assume that if there is an economic surplus, namely if $D_j > C_j$, the probability of an extension of the facility stock exceeds the probability of a reduction due to the profit expectation of the decision-makers.

The difference between (7.15) and (7.16) is that in (7.15) the decision function $\Phi_j^{(l)}(\mathbf{c})$ is assumed to be proportional to the economic surplus in zone j whereas in (7.16) a proportionality to the economic surplus per unit of the facility stock is considered. We further assume that the total cost of supply of facilities in j is proportional to the number of units of the facility stock:

$$C_j(t) = [k_j + r_j(t)]Z_j(t), \tag{7.17}$$

where k_j is a fixed constant and $r_j(t)$ is a measure for the rent of one unit of the facilities in zone j.

7.4 The Quasi-Deterministic Equations to the Dynamic Service Sector Model

By substitution of the transition rates (7.4), (7.10) and (7.11) into the master equation (7.1), we get the evolution in time of the probability distribution $P(\boldsymbol{c}, t)$, starting from a given initial distribution $P(\boldsymbol{c}, 0)$.

Equations of motion for the mean values (denoted by a bar) of the expenditure flows $\bar{T}_{ij}(t)$, the facility stock $\bar{Z}_j(t)$, the prices $\bar{p}_j(t)$, and land rent $\bar{r}_j(t)$ are obtained from the master equation according to Section 2.2.3. Therefore, we skip the derivation and proceed immediately to the quasi-deterministic equations of motion for our service sector model.

$$\begin{aligned}\dot{\bar{T}}_{ij}(t) = \varepsilon_1(t)\Big\{ &\sum_{k=1}^{L} \bar{T}_{ik}(t) \exp[v_{ij}(t) - v_{ik}(t)] \\ &- \sum_{k=1}^{L} \bar{T}_{ij}(t) \exp[v_{ik}(t) - v_{ij}(t)]\Big\},\end{aligned} \tag{7.18}$$

$$\dot{\bar{Z}}_j(t) = \varepsilon_2(t)\bar{Z}_j(t)\sinh\Big\{\lambda_2\Big[\sum_{i=1}^{L}\bar{T}_{ij}(t) - (k_j + \bar{r}_j(t))\bar{Z}_j(t)/Z_j(t)^{\delta}\Big]\Big\} \tag{7.19}$$

$$\dot{\bar{p}}_j(t) = \varepsilon_3(t)\bar{p}_j(t)\sinh\Big\{\lambda_3\Big[\sum_{i=1}^{L}\bar{T}_{ij}(t) - (k_j + \bar{r}_j(t))\bar{Z}_j(t)/Z_j(t)^{\delta}\Big]\Big\} \tag{7.20}$$

$$\dot{\bar{r}}_j(t) = \varepsilon_4(t)\bar{r}_j(t)\sinh\Big\{\lambda_4\Big[\sum_{i=1}^{L}\bar{T}_{ij}(t) - (k_j + \bar{r}_j(t))\bar{Z}_j(t)/Z_j(t)^{\delta}\Big]\Big\} \tag{7.21}$$

for $i, j = 1, 2, \ldots, L$ and $\delta = 0$ or 1.

Equations (7.18) - (7.21) constitute the explicit model. The utility functions $v_{ij}(t)$ will be specified in the next section. Then for given initial conditions $\{\bar{T}_{ij}(0), \bar{Z}_j(0), \bar{p}_j(0), \bar{r}_j(0)\}$ the trajectories for $t > 0$ can be computed. The hyperbolic sine function leads to an amplification of the reactions of the agents on economic disequilibrium. Near equilibrium (7.19) yields the Harris and Wilson[11] hypothesis (for constant prices and rents and assuming consumers in equilibrium).

By summing up (7.18) over j it can easily be seen that the total expenditure stock $\bar{E}_i$ of zone i remains constant with time:

$$\dot{\bar{E}}_i(t) = 0 \qquad \text{for } i = 1, 2, \ldots, L \tag{7.22}$$

and therefore satisfies a conservation law. From an economic point of view this is due to the fact that changes of residential location of individuals are not considered.

The estimation of the model parameters from empirical data can now be done by directly linking the transition rates (7.10), (7.11) to the corresponding empirical data (see Section 2.3.1 and Section 5.2.7).

The dynamic service sector model (7.18)-(7.21) comprises very different time scales. In the context of facility planning, we may often relate a leader/follower hierarchy to the relative speed of adjustment. For example, the time scale for altering the shopping area $\varepsilon_1(t)$, is certainly quite different from that of changing the facility stock $\varepsilon_2(t)$, the price dynamics $\varepsilon_3(t)$, and the rental dynamics $\varepsilon_4(t)$. Customers have no long-term commitments with the shops, and can easily make another choice every day. This very fast process can therefore be reasonably described in certain situations by its steady state. Thus, in retail problems, we may classify customers as followers with respect to retailers, and retailers as followers with respect to the owners of the centres. In order to provide planning authorities with more far-sighted suggestions for their policy instruments, it is recommended to denote them as leaders with respect to all the other actors in the model. This may lead to analytical simplifications of the model and is called adiabatic elimination (Haken[23]). Using this adiabatic elimination procedure the $\bar{T}_{ij}(t)$ are in equilibrium with the momentary values of $\bar{Z}_j(t)$, $\bar{p}_j(t)$ and $\bar{r}_j(t)$.

7.5 The Stationary Solution of the Service Sector Model

By considering the stationary versions of (7.18) - (7.21), we obviously obtain a complicated nonlinear transcendental system of equations for the stationary expenditure flows $\hat{T}_j$, the stationary scale of provision of facilities $\hat{Z}_j$, the prices $\hat{p}_j$ and rents $\hat{r}_j$. Stationary values are denoted by a hat.

It can be seen that the stationary version of (7.18) yields for the expenditure flows, which are denoted by $\hat{T}_{ij}$:

$$\hat{T}_{ij} = \hat{E}_i \frac{\exp(2\hat{v}_{ij})}{\sum_{k=1}^{L} \exp(2\hat{v}_{ik})} \tag{7.23}$$

Equation (7.23) can easily be proved by inspection. On the other hand the stationary solutions of (7.19)- (7.21):

$$\hat{D}_j = \hat{C}_j \tag{7.24}$$

must be fulfilled simultaneously. Therefore, there exists a coupling between the stationary demand $\hat{D}_j$ and the stationary values of the facility stock $\hat{Z}_j$ and the rent level $\hat{r}_j$ for each zone. In other words for each scale of provision and rent level there exists an appropriate demand.

By comparison of (7.23) with the special form of the flow pattern assumed by Harris and Wilson, Clark and Wilson[24] and Birkin and Wilson,[25] the corresponding utility functions belonging to these flows can be derived:

$$v_{ij}^{(W)} = \frac{1}{2}\Big[\alpha \ln Z_j(t) - \gamma \ln p_j(t) - \beta\, c_{ij}(t)\Big]. \tag{7.25}$$

In (7.25) we have assumed that the expenditure flows are always in equilibrium, or in other words that $\varepsilon_1 > \varepsilon_2, \varepsilon_3$ can be justified. Near equilibrium $|\Phi_j(\boldsymbol{c})| < 1$, and assuming (7.23), (7.25), the Harris and Wilson model is obtained (prices and rents are not considered):

$$\bar{T}_{ij}(t) = \frac{\bar{E}_i(t)\bar{Z}_j^{\alpha}(t)\exp[-\beta c_{ij}(t)]}{\sum_{k=1}^{L}\bar{Z}_k^{\alpha}(t)\exp[-\beta c_{ik}(t)]} \tag{7.26}$$

and

$$\dot{\bar{Z}}_j(t) = \varepsilon_2^*(t)\bar{Z}_j(t)\Big[\sum_{i=1}^{L}\bar{T}_{ij}(t) - k_j\bar{Z}_j(t)\Big]. \tag{7.27}$$

The parameter ε_2^* is a scaled time constant and different from ε_2. Therefore the application of our dynamic choice theory yields an alternative derivation of the spatial-interaction model, now based on microeconomic decision making and statistical averaging ideas. The chosen procedure has the advantage of introducing explicitly dynamic utility functions. However, the underlying assumptions are relatively weak compared to those of conventional economic choice theory.

From an economic point of view, however, it is not satisfactory that in the utility function (7.25) the prices of goods p_j and the transportation costs c_{ij} are treated in a quite different way (compare the term $\ln p_j$ with c_{ij}). Therefore, another assumption (Haag[20]) for the utility function is proposed, which seems to be more reasonable in this respect:

$$v_{ij}^{(H)} = \frac{1}{2}\Big[\alpha Z_j(t)[1 - Z_j(t)/Z_j^{sat}] - \gamma\, p_j(t) - \beta\, c_{ij}(t)\Big]. \tag{7.28}$$

This assumption (7.28) corresponds to a *Taylor* expansion of the utility function v_{ij} in state variables (compare Chapters 2, 5). A possible saturation of a zone j with respect to the facility stock Z_j is also taken into account. The saturation level of the facility stock is called Z_j^{sat}. Since both, p_j and c_{ij} are prices it is reasonable to put $\gamma = \beta$ (remark of *Andersson*). Then the utility function assumes the simple structure:

$$v_{ij}^{(A)} = \frac{1}{2}\Big[\alpha Z_j(t)[1 - Z_j(t)/Z_j^{sat}] - \beta\,(p_j(t) + c_{ij}(t))\Big]. \tag{7.29}$$

The utility $v_{ij}^{(A)}$ of consumers having residence in zone i and using facilities in zone j thus depends on the perceived prices of goods $(p_j + c_{ij})$.

From an economic point of view, it seems very plausible to assume that the shopping attitude of a consumer living in zone i will primarily depend on the offer of facilities in a certain retailing area j and his travelling costs c_{ij} from i to j. In comparison to these effects the shopping attitude of individuals living in any other residential area should not have an important influence and thus could be neglected in a first approximation. Therefore, Frankhauser[26] proposed for the utilities (prices and rents are not considered):

$$v_{ij}^{(F)} = \frac{1}{2}\Big[\ln[k_j Z_j(t)] - \beta\, c_{ij}(t) - \ln \tau_j(t)\Big] . \tag{7.30}$$

with

$$\tau_j(t) = \sum_{i=1}^{L} \exp(-\beta c_{ij}(t) . \tag{7.31}$$

The utilities (7.30) depend on the transport cost matrix in the same way as the utility functions (7.25), (7.28), (7.29). Because τ_j represents the sum over the transportation costs to reach a particular zone j, τ_j can be seen as a measure for the accessibility to reach the rest of the urban system from zone j. The utility function (7.30) seems to be similar to the utility assumption of Wilson for $\alpha = 1$. However, in (7.25) the offer in zone j is only introduced by the variable Z_j, usually interpreted as an arbitrary quantitative measure, e.g., the supply area. On the other hand, in (7.30) instead of Z_j the product $k_j Z_j$ is used, representing the cost of providing C_j. This is a well defined economic variable, for which economic data can be obtained.

Taking into account the meaning of the factor k_j (the cost per unit of providing Z_j), another interpretation becomes obvious, namely to understand k_j as a qualitative measure for the level of the quality and price standard in the area considered. Thus it can be expected that k_j will be higher in a city centre than in the suburbs. The product $k_j Z_j$ therefore includes qualitative as well as quantitative aspects.

By insertion of (7.30), (7.31) in (7.23) we obtain a set of decoupled stationary expenditure flows:

$$\hat{T}_{ij}^{(F)} = k_j \hat{Z}_j (\hat{\tau}_j)^{-1} \exp(-\beta c_{ij}) = \hat{D}_j (\hat{\tau}_j)^{-1} \exp(-\beta c_{ij}) \tag{7.32}$$

It can easily be seen that equation (7.32) is not only a correct solution of (7.23), but also fulfills the requirement (7.24).

It must be stressed that the L equations (7.24) are satisfied by (7.32) in an identical way. Thus there remain only the set of L^2 equations (7.32) to determine the stationary pattern ($\hat{Z}$, $\hat{T}$), containing $L^2 + L$ variables. Therefore L values could be chosen arbitrarily. Using this set, the L^2 remaining values then could be computed by (7.32). For example, an arbitrarily fixed set of $\hat{Z}_j$ could be inserted in (7.32) to calculate the L^2values T_{ij}. However, by summing up (7.32) the conservation law (7.22) for the expenditure distribution leads to an additional set of L conditions, which have to be satisfied also. Thus for given parameters β, c_{ij}, k_j, the closed set of equations lead to a unique set of stationary values $\hat{T}_{ij}$, $\hat{Z}_j$.

7.5.1 Relation between Expenditure Flows and Transportation Costs

The structure of the stationary expenditure flow pattern of (7.23) is highly influenced by the deterrence parameter β, or in other words by the transportation costs. For two interesting limiting cases the structure of the $\hat{T}_{ij}$ can be obtained even analytically:

case a)

$\beta \to 0$, the transportation costs can be neglected compared with other costs. Under this condition we find:

$$\lim_{\beta \to 0} \hat{T}_{ij} = \hat{E}_i \frac{\exp(2u_{ij})}{\sum_{k=1}^{L} \exp(2u_{ik})} \tag{7.33}$$

where:

$$u_{ij} = \lim_{\beta \to 0} v_{ij} \tag{7.34}$$

and $\hat{E}_i$ is the total expenditure stock in i, which we consider as a given constant according to (7.22). Obviously, $\beta \to 0$, will lead, in general, to high competition between different zones, since transport costs play no role. Therefore, many consumers are attracted to the city centre.

case b)

$\beta \to \infty$, transportation costs become an essential part of total costs. Then we get:

$$\lim_{\beta \to \infty} \hat{T}_{ij} = \hat{E}_i \, \delta_{ij} \, . \tag{7.35}$$

In this limiting case a quite different flow pattern can be observed. For high values of β ($\beta \to \infty$) the flows within each zone dominate, and a diagonal structure (7.35) appears. This will give small centres a good chance to survive.

To illustrate the influence of the deterrence parameter on the urban system by example, we consider a city centre (1) surrounded by three suburban zones (2), (3), (4) with the following properties: (see Fig. 7.3).

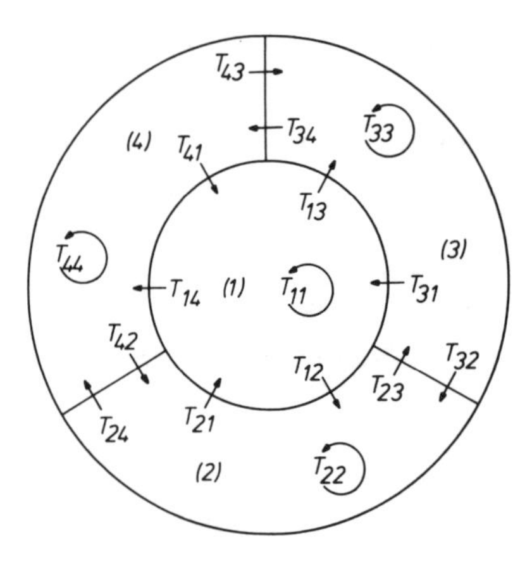

Figure 7.3: Representation of the urban system. (1) city centre; (2), (3), (4) different suburban zones

1. The transportation costs c_{ij} between neighbouring zones are assumed to be twice the costs c_{ii} within each zone
2. The facility stock Z_1 as well as the factor k_1 is three times higher in the centre than in the suburban zones, Z_j, k_j for $j = 2, 3, 4$
3. We consider the utility function (7.30)

In Fig. 7.4 a the expenditure flow distribution $\hat{T}_{ij}$ for $\beta = 0.35$ is depicted. The front row shows the flows $\hat{T}_{i1}$ into the city centre (1). In Fig, 7.4 b, the deterrence parameter is increased, $\beta = 70$. Competition and interaction between all zones characterize the first case, whereas a flow concentration occurs in the second case, in agreement with our considerations (7.33), (7.35).

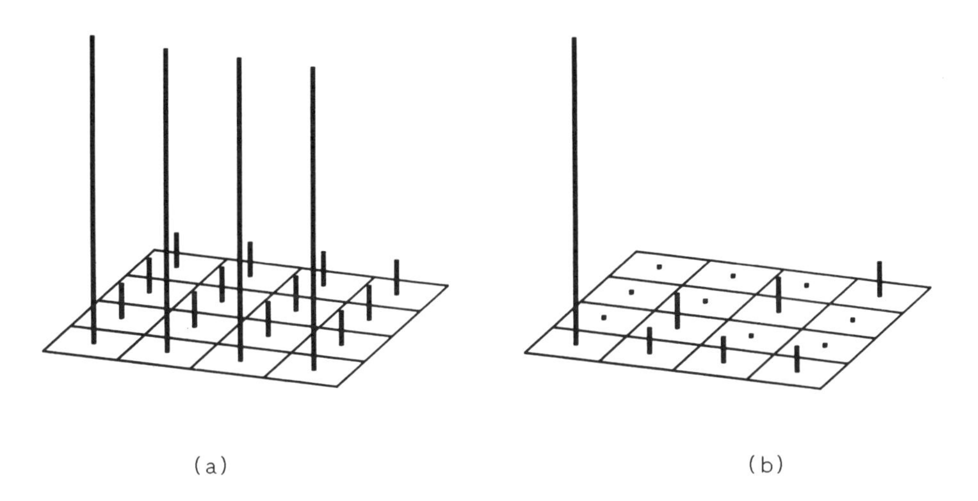

Figure 7.4: (a) Stationary expenditure flow distribution for $\beta = 0.35$
(b) Stationary expenditure flow distribution for $\beta = 70$

7.6 Dynamic Simulations and their Interpretation

To illustrate some dynamic aspects we consider the most simple version of our service sector model. This means we neglect prices and rents and assume the utility function (7.30). More detailed simulations and further remarks are given in Haag and Frankhauser.[27]

We start our simulation by assuming that the urban system is in its stationary state. The transportation cost matrix c_{ij} , and the k_j has been chosen as in Section 7.5.1, whereas a deterrence parameter β = 14 is now chosen, in order to assure an intermediate situation with respect to the influence of the transportation costs.

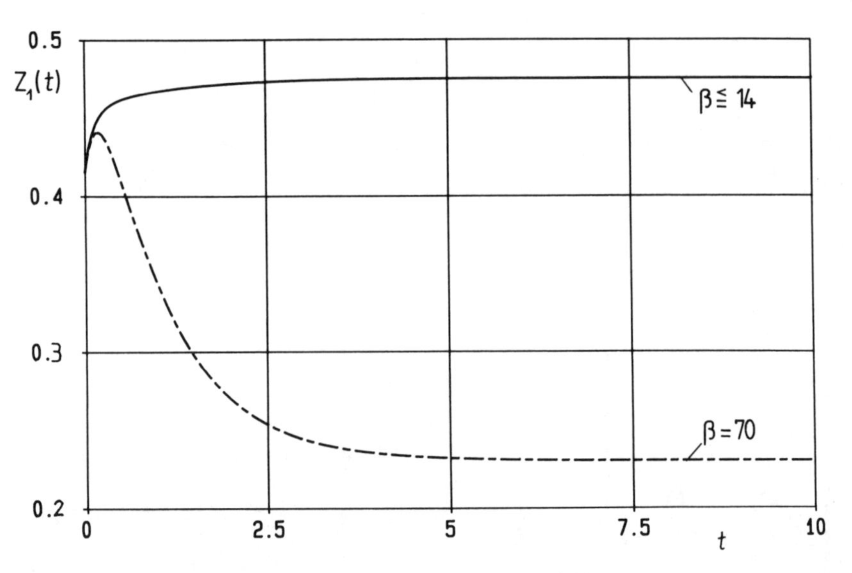

Figure 7.5 a: Scale of provision of facilities $Z_1(t)$ for β = 0.35, β = 14 (———), and β = 70 (·-·-·)

We shall now discuss a few numerical results. For instance we may ask for the response of the urban system to a sudden increase of the supply in a certain zone. To demonstrate this response, the arisal of new shopping facilities in the pheriphery zone (2) is simulated in Figures 7.5 a - 7.5 c for different values of β (β = 0.35 (- - -); β = 14 (———); β = 70 (- · - ·)).

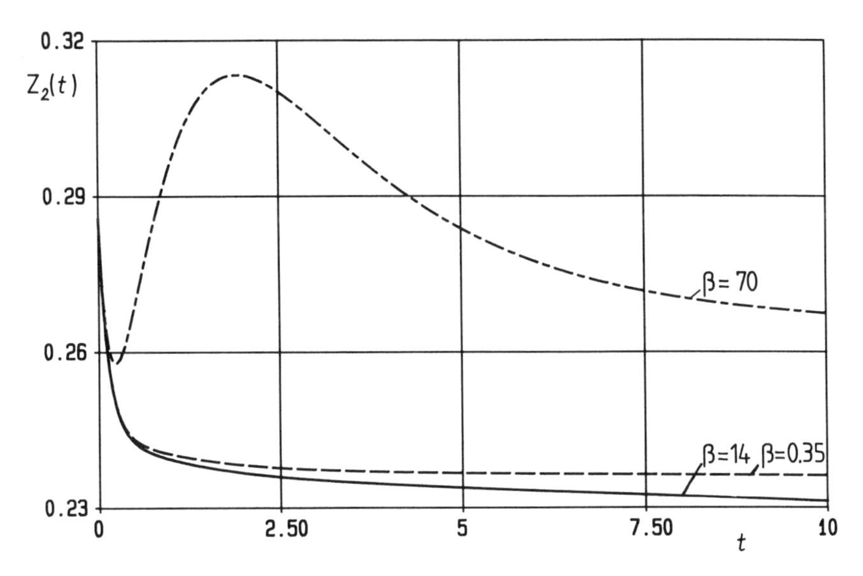

Figure 7.5 b: Scale of provision of facilities $Z_2(t)$ for β = 0.35 (----), β = 14 (———), and β = 70 (·-·-·)

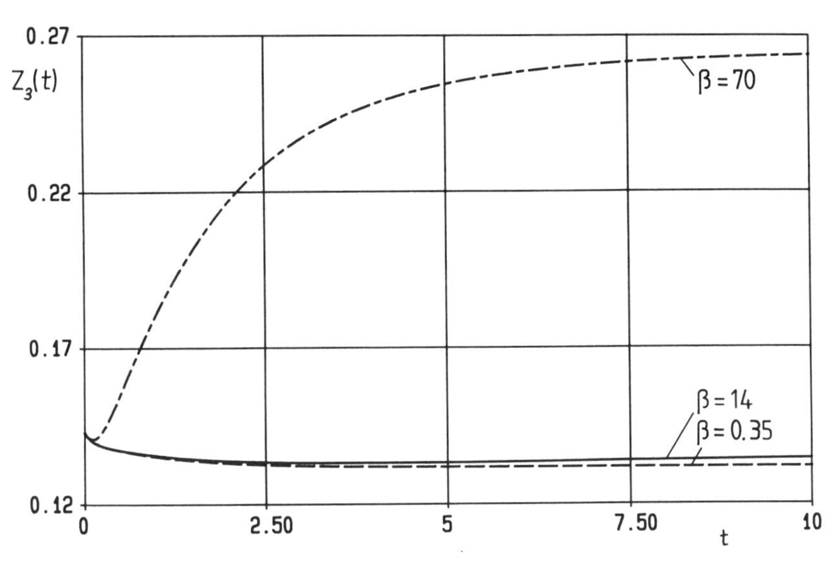

Figure 7.5 c : Scale of provision of facilities $Z_3(t)$ for β = 0.35 (----), β = 14 (———), and β = 70 (·-·-·)

7.6.1 The Influence of Low Transportation Costs

An increase of the starting level of supply $Z_2(0)$ cannot create a sufficiently high attractiveness of this shopping area, if β is small, thus leading to a decrease of supply in zone (2) for $t > 0$ (see Fig. 7.5b). The importance of zone (3) remains almost constant in this particular case. However, it must be emphasized, that the expansion of the facility stock in zone (2) has a positive impact on the demand-level in zone (1) and leads to an increase of the supply Z_1 in the time period under consideration. Thus the competition of service sector zones has some positive influence on the economic system.

The influence of low transportation costs becomes evident here: customers will not hesitate to go shopping in another zone, if β is small enough. This effect can also be observed by comparison of the expenditure flows. It can be seen that the big retail centres can survive even if new competition arises (at least if they are placed in the city centre) whereas small old shopping centres will lose their importance. Thus for this choice of parameters the model describes a situation of fierce competition calling for investors' activity.

7.6.2 The Influence of High Transportation Costs

Quite a different pattern is obtained when β is increased (-·-·-·). After a short time of prosperity the supply of zone (1) decreases in a very dramatic way (Fig. 7.5 a) to nearly half of its initial value, whereas zone (2) (Fig. 7.5 b) becomes stable at a rather high level after a small decrease. The evolution in zone (3) is surprising (Fig. 7.5 c). The increase of Z_2 leads to an unexpected high increase of the demand in zone (3) from 8.3% to 20.3% of the total demand (for the demand curves see Haag and Frankhauser). The supply Z_3 follows the evolution of the demand.

This clearly shows that high transportation costs - or alternatively - poorly developed transportation networks give small zones a good chance to survive, even without high investment activities. On the other hand, big shopping centres, which depend rather sensitively on customers coming from other residential areas are badly affected by such a change. Customers' trend towards shopping in their own residential area can also be well observed by considering the expenditure flow pattern.

Sudden changes on the demand and/or the supply side lead to overreactions and oscillations of the economic variables of the urban system. These effects are mainly caused by different speeds of adjustment of the supply and demand. In this particular case the initially doubled supply in zone (2) must again be augmented, when more consumers are attracted, but this large increase will diminish afterwards.

7.7 Concluding Comments

Starting from stochastic modelling of the decision behaviour of consumers, developers, retailers and landowners, a dynamic service sector model has been derived via our general theory starting from individual decision processes. Therefore, a microeconomic foundation has been provided for the macroeconomic equations of motion for the expenditure flows, the facility stock, the prices - and rent dynamics. The link thus provided enables us - in analogy to Chapter 2 - to establish a general estimation procedure for all parameters of the model, even if the service sector is out of equilibrium. This aspect is very important if the model is to be used for planning purposes.

By introducing some additional assumptions (e.g., relating to the decision behaviour of the different agents as well as for the time scaling parameters (adiabatic elimination procedure)) a slightly extended but conventional spatial interaction model for the expenditure flows can be obtained whereas the dynamics of the facility stock reproduce *Wilson's* extension of the *Harris and Wilson* model. On the other hand in our model frame new microeconomic data collected, e.g., by inquiry of agents could modify the utility and profit functions and perhaps lead to a modified and even more realistic model.

Summarizing the numerical simulations (of the most simple version) of our service sector model we have confirmed, that a decrease of transportation costs leads to a pronounced competition between the zones. In the short run, the propensity of investors in the zone of smaller facility stock to increase their supply has a positive effect on the demand attracted. This investors' behaviour is directed towards the equilibrium situation of the spatial urban system and thus accelerates this adjustment process. Although it is true that the final stationary state is independent of initial changes in the scale of provision, the long relaxation time of spatial urban systems in general prevents the system from approaching its stationary state. Since parameters involved in socio-economic models can only be assumed as constant for moderately long time intervals, exogeneous changes of model parameters like the change of the infrastructure may lead to different supply and demand patterns in the meantime before the system reaches equilibrium. Thus the value of models like those discussed here seems to lie in short term or medium term forecasting.

Chapter 8

Further Applications and Extensions

In the previous chapters we studied different cases of individual choice behaviour. We have seen how decisions of individual agents may result in a tremendous growth of settlements (Chapter 3), how intra- urban and inter- urban migration processes may lead to migratory phase transitions and even chaotic trajectories (Chapters 4, 5 and 6) and how the supply side and demand side interact in a simple service sector model (Chapter 7). In so far, we tried to incorporate - starting from the most simple migratory process - more and more economic and social effects, in order to capture with our dynamic theory of decision processes realistic situations in certain sectors of society.

We shall now briefly discuss further possible applications and give an outlook on current research in this field.

8.1 Knowledge, Innovation, Productivity

In recent years the world economy has been subject to major shocks and structural changes. Examples can easily be found: the combination of inflation and increasing levels of unemployment (*stagflation*), large reductions of construction and investment activities, excess capacities in steel and coal, reversals of urbanization trends in many developed countries. An extensive review of the impact of these processes on spatial dynamics can be found in Nijkamp and Schubert.[1]

Technological Research and Development (R&D) has been assumed by many economists to provide a solution to the worldwide slowdown of economic growth.

Traditionally economists have treated the stock of knowledge and its technological counterpart, R&D, as an exogeneous factor influencing the growth possibilities of an economy. In some macroeconomic growth models technological progress therefore is only treated as a shift parameter. However, it seems more fruitful to deal with the question how knowledge creation interacts with standard capital formation in an economy (see Andersson[2]).

In this context we follow Andersson's argumentation by introducing the stock of knowledge $G(t)$ as a variable and the change of it as $\dot{G}(t) \equiv R(t)$, where $R(t)$ indicates the volume of R&D. The capital stock at time t is denoted by $K(t)$. Knowledge is looked upon as a public good. This means that knowledge participates in the production process as an input without being used up by the process. Being a public good, knowledge can be used without conflict between all commodity producing activities in a sufficiently small region. This implies that disaggregation of the decisions on production and allocation of recourses becomes critical for aggregate efficiency.

The regional dimension is extremely important in all studies of public goods. Since the availability of technological information is evidently much better on smaller than on larger distances accessibility is a reasonable way of representing the economic effects of a spatial distribution of a public good like technological knowledge.

8.1.1 Knowledge as an Endogeneous Input of the Growth Process

We consider a spatial economy consisting of L regions. The capital stock available in region i (i = 1, 2, ..., L) is denoted by $K_i(t)$, the stock of knowledge by $G_i(t)$. We assign to each region i a production function (neo-classical production function):

$$Q_i(\boldsymbol{K}, \boldsymbol{G}, \dots) = \left[c_1 K_i(t)^\rho + c_2 \tilde{G}_i(t)^\rho \right]^{\nu/\rho}, \tag{8.1}$$

where we have introduced as a fairly general concept of accessibility of knowledge $\tilde{G}_i(t)$:

$$\tilde{G}_i(t) = \sum_{j=1}^{L} f_{ij} G_j(t), \tag{8.2}$$

where $f_{ij} = f_{ji} \geq 0$ is a semi-definite deterrence function (compare Chapters 2, 5). A fraction of the production, $s_i(t)Q_i(t)$, is used as non- material investment. Therefore, $s_i(t) \geq 0$, indicates the total savings or investment ratio in region i at time t. We subdivide the non-material investment into one part, $\mu_i(t)s_i(t)Q_i(t)$ which is used for knowledge production (volume of R&D), where $\mu_i(t)$ is the research investment in region i and the part, $(1 - \mu_i(t))s_i(t)Q_i(t)$, which is used for capital production.

A reasonable assumption for the dynamic equations of the capital stock $K_i(t)$ and the stock of knowledge $G_i(t)$ reads:

$$\dot{K}_i(t) = (1 - \mu_i(t))s_i(t)Q_i(\mathbf{K}, \mathbf{G}) - \delta_K K_i(t) \tag{8.3}$$

$$\dot{G}_i(t) = g_i(t)\mu_i(t)s_i(t)Q_i(\mathbf{K}, \mathbf{G}) - \delta_G G_i(t) \tag{8.4}$$

where $g_i(t) \geq 0$ is the productivity of the R&D- producing sector, and δ_K, δ_G are the rates of decrease of the capital stock and the stock of knowledge, respectively.

Numerical s imulations with this model (8.3), (8.4) have shown (Andersson and Mantsinen[3]) that it is relatively stable for any set of neo-classical production function. An interesting property of this kind of models is their tendency to transform spatial distances into time lags. Growth rates of the periphery regions thus generally increase at a later stage than growth rates of centrally located regions.

However, in addition to the above mentioned model, we (Andersson and Haag[4]) want to formulate in terms of our theory of decision processes the investment ratio $s_i(t)$, as well as the share of investment going into R&D, $\mu_i(t)$.

8.1.2 Regional Decisions about the Investment Ratio

As we are dealing with a production function, we always assume one leading firm in each region i. Without this simplifying assumption the equations of motion for the capital stock (8.3) and the stock of knowledge (8.4) have to be split and the production functions of each of the different firms have to be considered. This generalization can be done without specific difficulties.

The decision process is again a stochastic process, where we are interested in the probability to find a certain spatial decision pattern for the investment ratio $s_i(t)$. The individual transition rates (we consider the single firm as agent) for an increase or decrease of the investment ratio by one unit are again considered formally as birth/death rates.

It is reasonable to assume that the transition rates $w_{i+}^{(s)}$, $w_{i-}^{(s)}$, for an increase or a decrease of s_i will be influenced by the investment ratios s_j of the other regions as well. The spatial influence of the other regions is taken into account via the deterrence function f_{ij}. Then a rather general formulation for the total transition rates which is flexible enough reads:

$$w_{i\pm}^{(s)} = \varepsilon_s \sum_{j=1}^{L} f_{ij}\, s_j\, e^{\pm \Phi_{ij}} > 0\,, \quad i = 1, 2, \ldots, L \tag{8.5}$$

for rather general functions Φ_{ij}. The factor ε_s describes the speed of adjustment. Since birth/death rates must not be negative, the condition $\varepsilon_s > 0$ has to be fulfilled. Spatio-economic considerations about the decisions of firms with respect to the investment ratio are described by the decision function Φ_{ij}.

We assume that the probability for an increase of the percentage share of total investment exceeds the probability for an increase of s_j in region j, if:

$$\frac{\partial \pi_i}{\partial K_i} > \frac{\partial \pi_j}{\partial K_j}\,, \tag{8.6}$$

where $\partial \pi_i / \partial K_i$ can be denoted as *marginal profitability of capital.*

The profit is given by:

$$\pi_i = Q_i - r_i K_i , \tag{8.7}$$

where r_i is the rate of interest in region i. Therefore, a reasonable assumption for the decision function Φ_{ij} is given by:

$$\Phi_{ij} = \lambda_s \left(\frac{\partial \pi_i}{\partial K_i} - \frac{\partial \pi_j}{\partial K_j} \right) \tag{8.8}$$

where the parameter λ_s describes the intensity of the response due to differences in regional marginal profitabilities.

8.1.3 Decision Processes Concerning the Percentage Share of Research Investment

One of the most important questions in the discussion of technology and structural change is related to the problem of an optimal choice of the research investment $\mu_i(t)$ as a percentage share of total invest- ment. In this section we will briefly discuss two approaches:

a) manager controlled decisions about $\mu_i(t)$
b) $\mu_i(t)$ fixed via a political decision process

case a)

Let us assume that the leading firms of each region make their own decisions about their percentage share of non-material investments going into trainee programs (R&D) for their own employees. We assume that the transition rate for an increase or decrease of $\mu_i(t)$ can be modelled via:

$$w_{i\pm}^{(\mu)} = \varepsilon_\mu \sum_{j=1}^{L} f_{ij} \mu_j e^{\pm \Psi_{ij}} > 0 , \quad i = 1, 2, \ldots, L \tag{8.9}$$

where $\varepsilon_\mu > 0$ is a time scaling factor and function Ψ_{ij} takes into account the economic advantage (disadvantage) to invest R&D in region *i* instead of region *j*. We assume that this decision process is profit oriented. Then a reasonable assumption is:

$$\Psi_{ij} = \lambda_\mu \left[\left(\frac{\partial \pi_i}{\partial \tilde{G}_i} - \frac{\partial \pi_i}{\partial K_i} \right) - \left(\frac{\partial \pi_j}{\partial \tilde{G}_j} - \frac{\partial \pi_j}{\partial K_j} \right) \right] \tag{8.10}$$

where the parameter λ_μ describes the response to differences in profitabilities.

case b)

The role of a political decision process can be seen as providing an optimal framework for the social and economic development of a region. The choice of a certain rate of research and development must be made in cooperation by the users of knowledge. The most common way of acting collectively is taxation (collective financing of the public good *knowledge*). Therefore, the problem of an optimal taxation μ_i arises (see Andersson and Haag).

Instead of (8.9) we assume that under this condition the transition rates of μ_i are now given by:

$$w_{i\pm}^{(\mu)} = \varepsilon_\mu \mu_i \, e^{\pm \Psi_i} > 0, \quad i = 1, 2, \ldots, L \tag{8.11}$$

where the political decision function Ψ_i is assumed as:

$$\Psi_i = \lambda_\mu \left[\sum_{j=1}^{L} \frac{\partial \pi_j}{\partial G_i} - \frac{\partial \pi_i}{\partial K_i} \right], \tag{8.12}$$

where $\partial \pi_j / \partial G_i$ is the marginal profitability of knowledge capital invested in region *i* accruing to region *j*.

8.1.4 Some Conjectures

The dynamic equations of motion (mean value equations) for the investment ratio $\bar{s}_i(t)$, and the share of research investment $\bar{\mu}_i(t)$ are obtained by insertion of (8.5) and (8.9) or (8.11) in:

$$\dot{x}_i(t) = w_{i+}^{(x)}(\boldsymbol{x}) - w_{i-}^{(x)}(\boldsymbol{x}) \qquad \text{for } i = 1, 2, \ldots, L \quad \textit{and} \quad x_i = \bar{s}_i, \bar{\mu}_i \tag{8.13}$$

The different versions of (8.13) together with (8.3), (8.4) constitute the dynamic system of equations. The stationary values of the capital stock and the stock of knowledge are determined by a nonlinear system of transcendental equations. It can be proved that the stationary solution does *not* coincide with the profit-maximizing solution:

$$\max_{\{\boldsymbol{K},\boldsymbol{G}\}} \pi_i(\boldsymbol{K}, \boldsymbol{G}, \boldsymbol{\mu}, \boldsymbol{s}) \tag{8.14}$$

because of $\partial\pi_i/\partial K_i = 0$ and $\partial\pi_i/\partial G_i = 0$ the system of stationary equations is not satisfied.

The difference between the profit maximizing solution and the *realized* stationary solution may create innovative steps at certain frontiers to come closer to the profit maximizing solution. On the other hand this disturbes the stationary conditions of the other regions and the whole system starts to move. Therefore, the existence of economic regular or chaotic cycles seems to be inherent in spatial dynamic systems.

At the end of this section we shall us discuss a few numerical simulations. For simplicity only 3 regions have been considered. The parameters used in the simulation are listed in Table 8.1. The results shown in Figures 8.1 to 8.3 correspond to the set of dynamic equations (8.3), (8.4), and assume a politically controlled process for the percentage share of research investment (8.11), (8.12).

Table 8.1

Trend parameters used in the numerical simulation of Figures 8.1 -3

λ_s	λ_μ	g_1	$g_2=g_3$	$\delta_K=\delta_G$	$c_1=c_2$	ρ	λ	ε_s	ε_μ	$r_1=r_2=r_3$	Figure
10	1	1	0.5	1	1	0.5	1	0.1	0.05	2 .0	8.1
10	1	1	0.5	1	1	0.5	1	0.1	0.05	2 .5	8.2
10	1	1	0.5	1	1	0.5	1	0.1	0.05	3.0	8.3

In Figures 8.1 - 8.3 we have plotted normalized variables. An increase of the rate of interest from a low level to a medium level seems to stabilize the economy and to accelerate its movement towards a stationary solution. However, if r_i exceeds a critical value the solution, which appeared stable at first, becomes unstable and a catastrophic phase transition sets in.

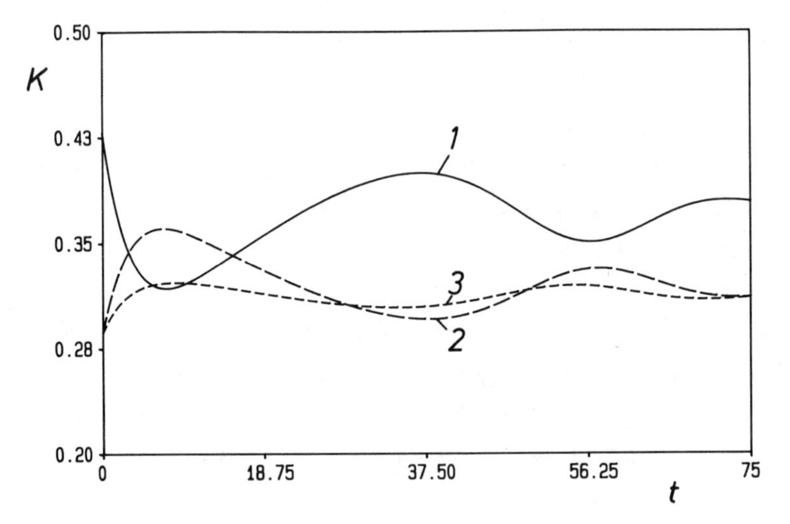

Figure 8.1 a: The capital stock K_i (t). Scaled rate of interest r_i = 2.0

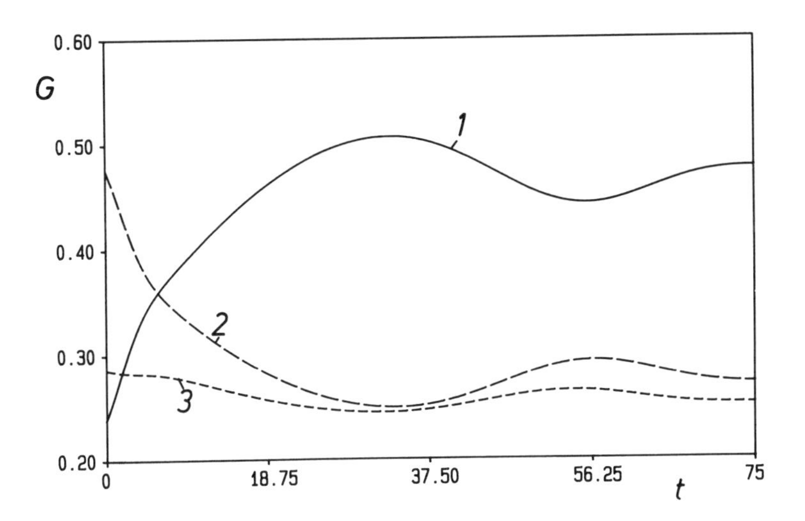

Figure 8.1 b : The stock of knowledge $G_i(t)$. Scaled rate of interest $r_i = 2.0$

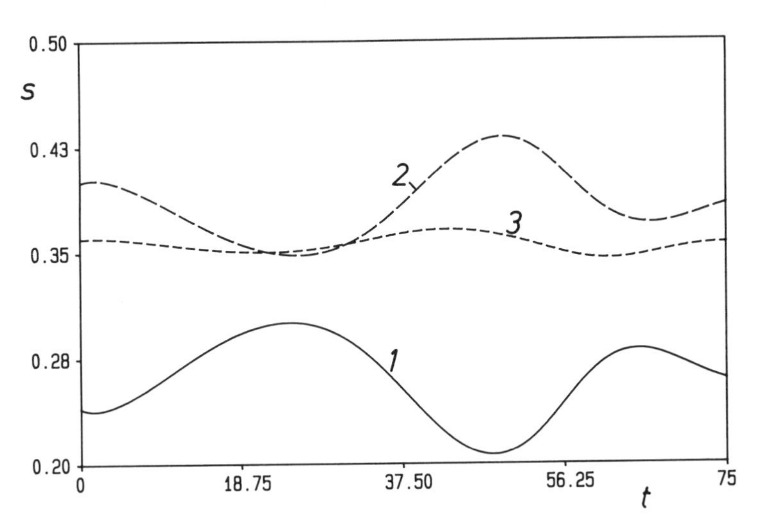

Figure 8.1 c : The investment ration $s_i(t)$. Scaled rate of interest $r_i = 2.0$

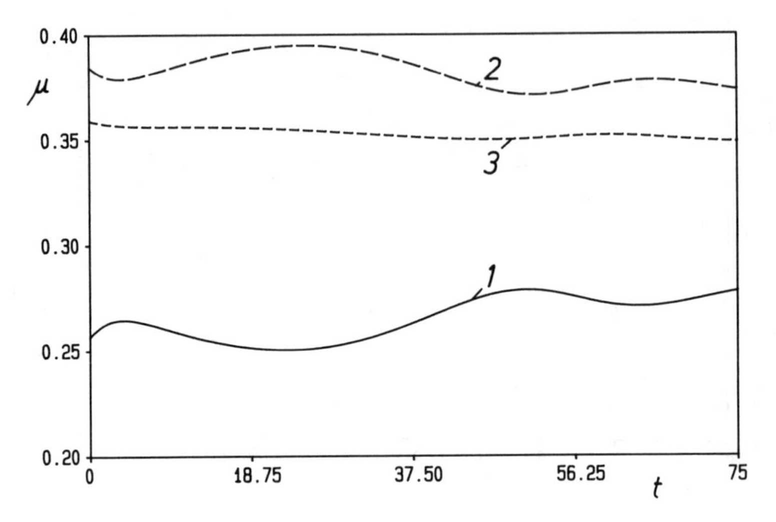

Figure 8.1 d: The research investment $\mu_i(t)$. Scaled rate of interest r_i = 2.0

In Fig. 8.1 we observe that because of the higher productivity of the R&D-producing sector $g_1 > g_2 = g_3$, in the long run region 1 (——)

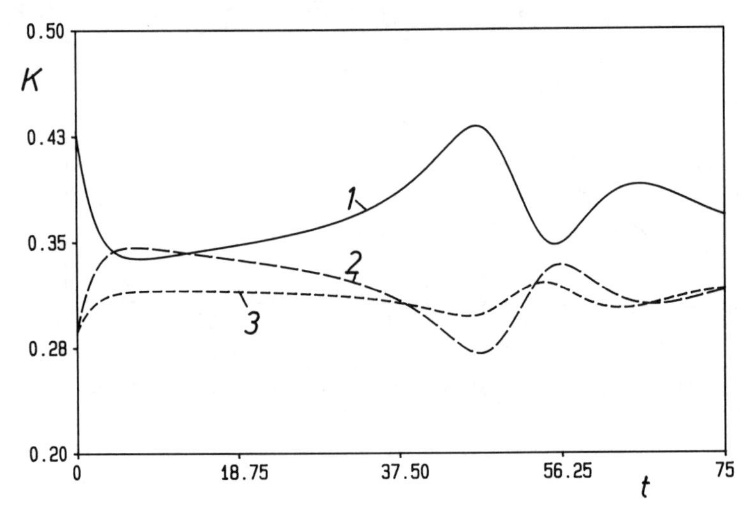

Figure 8.2 a: The capital stock $K_i(t)$. Scaled rate of interest r_i = 2.5

is favoured (high levels of the capital stock and the stock of knowledge and simultaneously low levels of the rate of taxation and the investment ratio compared with the other two regions). Furthermore, the economy tends to economic oscillations.

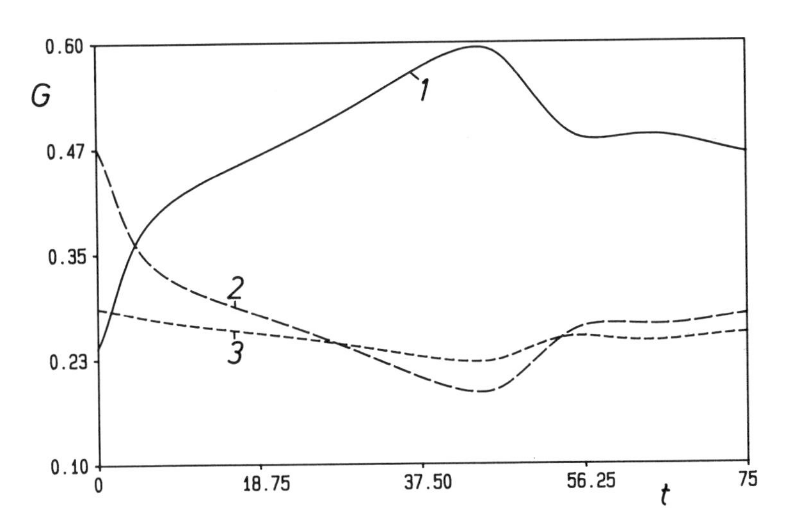

Figure 8.2 b: The stock of knowledge $G_i(t)$. Scaled rate of interest $r_i = 2.5$

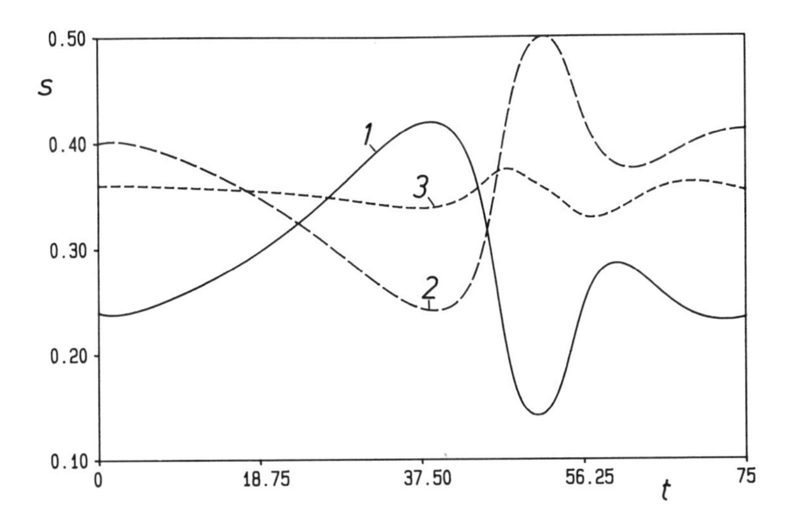

Figure 8.2 c: The investment ration $s_i(t)$. Scaled rate of interest $r_i = 2.5$

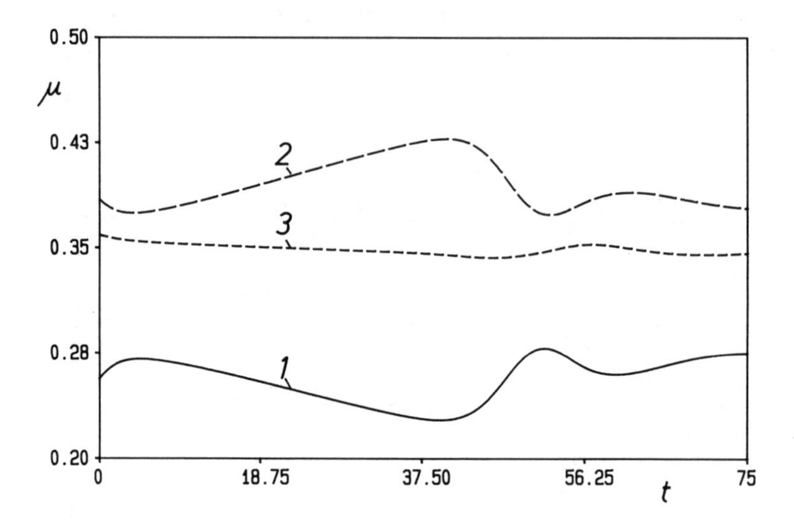

Figure 8.2 d: The research investment $\mu_i(t)$. Scaled rate of interest $r_i = 2.0$

A further increase of the rate of interest (see Fig. 8.2) enhances the range of the economic variations. Especially the evolution of the investment ratio $s_i(t)$ is characterized by high uncertainties.

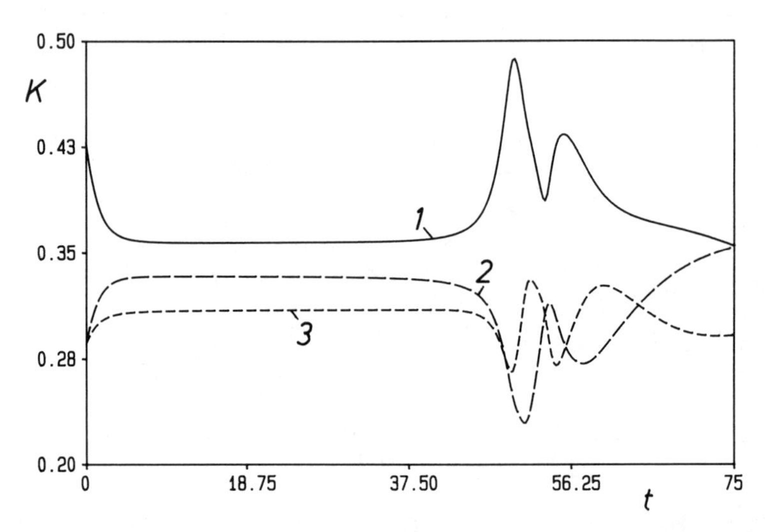

Figure 8.3 a: The capital stock $K_i(t)$. Scaled rate of interest $r_i = 3.0$

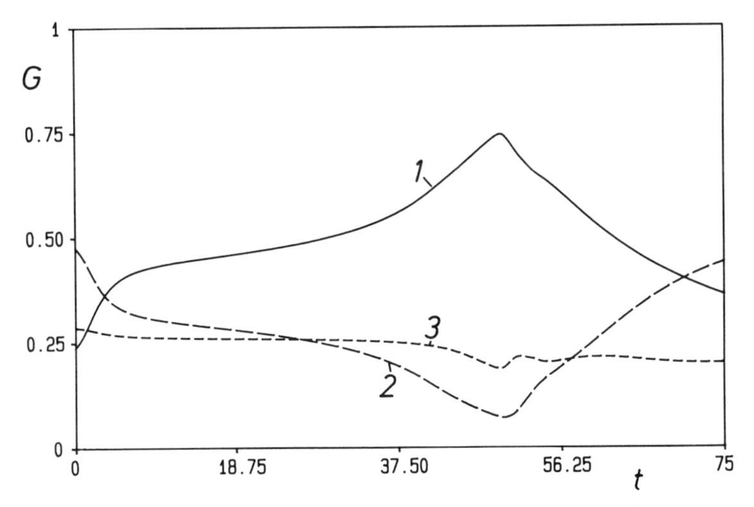

Figure 8.3b: The stock of knowledge $G_i(t)$. Scaled rate of interest $r_i = 3.0$

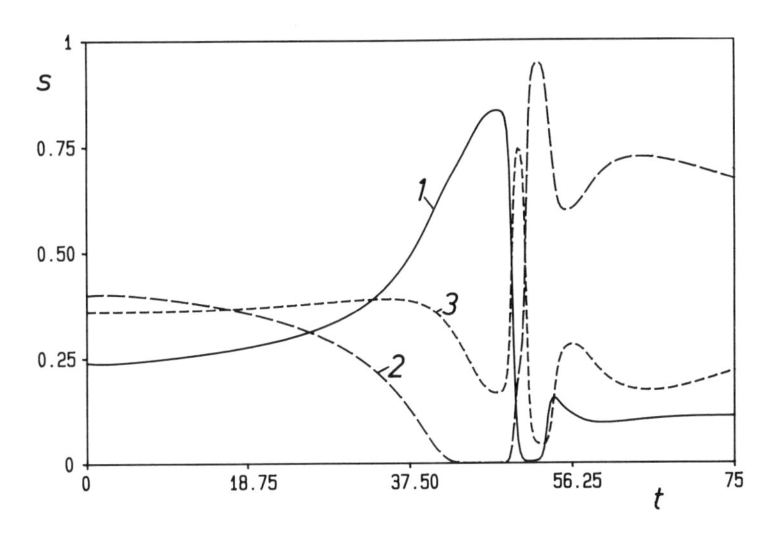

Figure 8.3 c: The investment ration $s_i(t)$. Scaled rate of interest $r_i = 3.0$

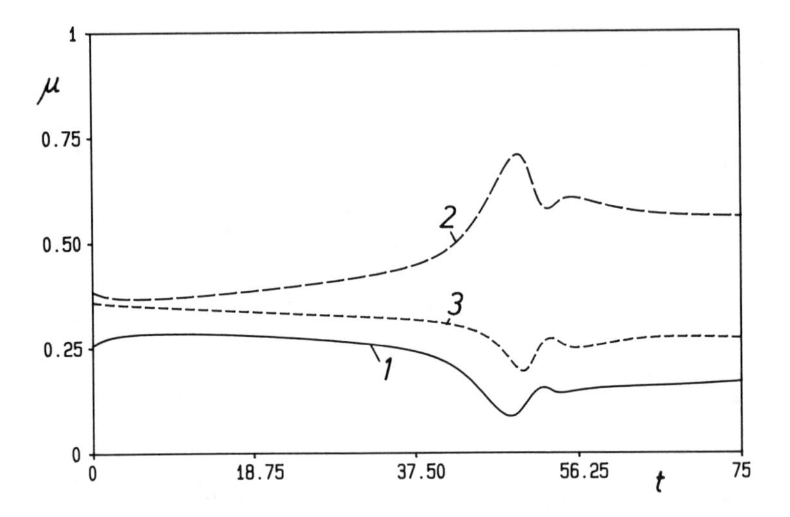

Figure 8.3d: The research investment $\mu_i(t)$. Scaled rate of interest $r_i = 3.0$

However, in Fig. 8.3 the economy seems to evolve in a rather stable and predictable way for a relatively long time period. In order to keep its first position, the investment ratio of region 1 has to be continuously increased. A catastrophic breakdown of the economy sets in leading afterwards to a completely different spatio- economic pattern. For more detailed information see Andersson and Haag.

The strategy of firms to invest in regions with a higher marginal profitability as well as the politically motivated taxation for financing of the R&D sector thus may lead to cyclical variations and instabilities of the economic system.

Generally speaking, especially if a system is driven by mutually conflicting forces such as e.g., differences in short run and long run adjustments, different interest groups, etc., the behaviour of normally continuous and stable trajectories may exhibit various perturbations including shocks.

8.2 Economic Cycles

The economic nonequilibrium process may be viewed as a superposition of long- and short-term cycles. This is a mechanistic concept. The microeconomic reasons for the appearance of these cycles are in principle very different.

8.2.1 Short-Term Cycles

Short-term economic cycles, with periods of only a few years, may arise within a quasi-stationary phase of a long-term cycle. The economic evolution in that phase is viewed as driven by an ensemble of products, and market growth is promoted by competing entrepreneurs who are capitalizing on the driving forces of technological progress.

A description of these business cycles in terms of our dynamic theory of decision processes is given in Weidlich and Haag[5] and Haag et al[6] The model is applied to changes in industrial strategic investment in the Federal Republic of Germany between 1956 and 1978. A further interesting application of this model for energy investment in the U.S. manufactoring sector is given by Pena-Taveras and Cambell.[7]

Subsequently we shall follow the argumentation in Weidlich, Haag and Haag et al. We assume that the decisions of interacting entrepreneurs about the share of expansionary versus rationalizing investment are readjusted under the pressure of the market:

If the majority of investors tends to maximize profits at a given point in time by expanding their business operations, then some innovators or pioneers will try to improve their market position by adopting a non-conformist strategy in an attempt to capture quasi-rents due to differentiation. During an upswing characterized by expansionary investments undertaken by a majority of investors, these trend setters tend to redirect their efforts and start pushing back the cost frontier by means of cost reducing investments. They thereby force others to imitate and also to undertake rationalizing investments in the expectation of further cost reductions. At other times when a downswing is well under way due to the contractionary effects of rationalizing investments undertaken by a majority of investors, the trend setters start moving towards the quality section of the best practice frontier, introducing better products and implementing investment plans for expanding facilities. Thereby others are forced to imitate this expansionary and quality updating behaviour, thus creating the synchronization to be observed in the occurance of business cycles.

The model which takes into account these decision readjustment processes indeed leads to limit cycle solutions under appropriately chosen trend parameters and thus is able to simulate the short term business cycle behaviour.

8.2.2 Long-Term Cycles

Long term economic cycles, if they exist, are expected to arise for other, deeplying structural reasons. The main *phases* of long-term cycles such as *recovery*, *prosperity*, *recession* and *depression* have been analyzed in detail by Kondratief,[8] Schumpeter [9] and many others. The long-wave debate (Bianchi et al,[10] Vasko[11]) points to the variety of theoretical models of the long wave, but also to the need for an integrated theory. Two main questions have been discussed extensively in the last years:

1. How can a recession occur at a time when the factors of production are abundantly available ?
2. Do long-cycles really exist or are they the outcome of our limited data base ?

After the collection of empirical observations (at least four *Kondratief-cycles* could be identified) and after careful checks the majority of researchers seem to believe in their existence. However, the driving mechanisms are rather unexplained and open for competing discussions. Nevertheless, as Delbeke[12] pointed out: . . . *three main categories of theory can be discerned: the real, the monetary, and the institutional.* Delbeke classifies the real theories according to the production factor which is supposed to be crucial.

. . . *All three types of approach agree, implicitly or explicitly, that the long wave is inherently based on capital accumulation and is therefore most noticeable in the industrial economies, especially the market oriented economies. Moreover, it seems that a fruitful integration of long-wave theories is not possible if these three main categories are not included.*

Because the role of entrepreneurship is considered to provide the most important approach (Bruckmann[13]) and *Schumpeterian* innovation theories are certainly the most widely debated onces since *Das Technologische Patt* (Mensch[14]), entrepreneurial decision processes of all kinds are certainly fundamental and may be viewed as a good starting point for an integrated approach. Therefore, it seems to be reasonable to apply our dynamic theory of decision processes to this interesting problem. A first outline of a formal theory of long-term economic cycles is given by Mensch et al.[15]

An innovative impulse can be created by a clustering of *basic innovations* (Mensch[16]). These clustering processes of innovations may be one of the main forces leading to long waves in the economic development (Bianchi et al[17]). In general, these clustering trends are both temporal and spatial events. Spatial clustering may lead to growth pools and growth centres in the economic system. A spatial and temporal innovation impulse diffuses through the spatial system and penetrates the multisector economy (Fischer,[18] Wiseman[19]). The labour market, the distribution of income between wages and profits, the attractiveness of a residential area and the service sector are influenced by such an event. Therefore, innovations contribute to variations on almost all levels of the socio-economic system.

8.3 Housing and Labour Market

Recently, the complex relationship between technological change, employment and regional growth has attracted wide attention. Especially the last stagnation has induced many research efforts among economists in this area. Increasing attention is called for labour market aspects of long term industrial dynamics (Forrester[20]).

Despite the world wide importance of labour market dynamics it is still an unresolved question whether innovation favours employment or has totally adverse impacts. Therefore, intense theoretical and empirical research on labour saving and labour augmenting technology is needed.

It is increasingly questioned whether conventional macro-oriented economic approaches are still sufficient and whether disaggregated behavioural approaches at the firm level could not be able to provide a better analysis framework (Nijkamp[21]). Clearly, the relationship between technological change and employment is a complex one. The impact of innovations on firms, sectors, cities or regions is of high interest for urban and regional industrial and R&D policy. The political tools must be chosen carefully in order to prelude future employment effects.

The locational requirements of new firms, R&D centres or new administrative centres and their positive or negative influence on economic growth focus our attention on the regional and urban dimensions of industrial development and related labour market conditions. The observed sectorial shifts in industrial growth patterns in many countries appear to be accompanied by drastic shifts in the role of urban agglomerations. It is worth noting, however, that the changing role of cities and regions in a spatial system leads to new evolving spatial configurations. Empirical investigations of migration flows, residential and labour mobility, the spatial location pattern of new firms confirm this result. Various large metropolitan areas tend to lose their innovative potential in favour of medium sized cities (Pumain and Saint-Julien[22]). The theory of job search (Rogerson,[23] Lippman and MacCall[24]) and residential search (Clark[25]) is concerned with optimal behaviour of individuals to maximize their utility. Optimal policies for firms are concerned with maximizing profits (Mortensen,[26] Eaton and Watts[27]).

Of course, this interesting field of research gives rise to a large amount of potential applications of this dynamic theory of decision processes. For instance, closely connected to the model of Chapter 5 is the consideration of residential and labour mobility in Haag[28-29] which is briefly reviewed in the following.

Regional labour market conditions depend in a complex manner on the decisions of *individuals* on the supply side and *firms* on the demand side of the market. Changes in social and technological conditions have a profound impact on the structure and mechanism of these labour markets. The individuals must decide whether to quit or retain their present jobs, to search for new employment, perhaps in another region, and/or to search for a new residential location which increases their utility. Firms must set desired employment levels and determine the number of job opportunities in the different regions, taking into account the market conditions. These highly complex interactions between the labour market and the housing market require an adequate explanation of the nature of the decisions made by the different agents.

Housing consumption decisions are also discrete in nature. Choices are made among a household specific choice set to take into account for institutional, informational and income-based constraints of the choice process. Government subsidies (such as housing and rent allowances, non-interest-bearing state loans) must also be considered (for details and further references see Fischer and Aufhauser,[30] Maier and Fischer[31]). Thus, a discrete choice model is again an appropriate starting point.

Multinomial logit models are often used for the analysis of the structure of the market (Rouwendale [32]). It is shown in the appendix (Chapter 9.7) that the multinomial logit model can be embedded into the framework of this book. However, the metropolitan housing market is in general not in its equilibrium state, since political decisions and economic and social imbalances always create migratory stress. Hence, a more sophisticated treatment of the individual decision processes of all agents of the system is required.

8.4 Concluding Remarks

Concepts of a dynamic theory of decision processes have been presented. This broader overall framework with full probabilistic dynamic underpinning also facilitates extensions in a number of directions. For instance, it offers a new way to link the microeconomic decisions of a group of individuals to the dynamic behaviour of a few but collective macrovariables. A new approach to parameter estimation has also been introduced, and it has become obvious how synergetic effects (agglomeration effects) in decision processes as well as the complicated interactions among different agents can be treated in a stochastic framework. Furthermore, it became obvious in Chapter 3 why the equilibrium of the macrostate generally does not coincide with microeconomic optimality. Exogeneous macrovariables may become endogeneous when it is possible to define an underlying decision process (on the microlevel) or in other words, when a decision-maker can be identified. Then the configuration space of the master equation must be extended accordingly and via the master equation a corresponding set of (in general non linear) mean value equations can be derived. The originally exogeneous macro variable is then replaced by an endogeneous macro variable. This has been demonstrated by means of examples in Chapters 4, 7, and 8.

In the interesting field of migratory processes phase transitions from an ordered state into a chaotic migration pattern may happen (Chapters 5, 6). Deterministic chaos requires at least three independent dynamic macro variables. Interacting subpopulations with non linear transition rates showing positive intra-group interactions and at least between two subpopulations repulsive inter-group interactions of different strength fulfil the necessary conditions for the occurrence of chaotic trajectories. However, in most of real world applications we expect a regular behaviour of the flows, since the estimated parameters lie considerably outside the critical parameter range.

The microfoundation of a class of spatial interaction models provides an interesting insight into the structure of model building via this stochastic approach (Chapter 7). The inter-related choice processes of different agents are reflected on a specific structure of the dynamic equations of the macro variables. It became obvious how extensions in different directions of the discussed models can be performed towards an integrated urban theory.

A briefly outlined application of this dynamic decision model on innovation processes, where we consider knowledge as an endogeneous input of the growth process is given in Chapter 8. An interesting aspect of this analysis is that the profit maximizing solution does not coincide with the stationary solution of the spatial system under consideration. The nested dynamics of the housing and labour market is briefly mentioned. First applications of this theory, using the concept of household specific choice sets to account for institutional, information and income-based constraints of the choice process, are published. The spatial aspects of the residential and labour mobility processes are crucial for a better understanding of regional growth or decline. In all these considerations it seems to be increasingly risky to assume equilibrium conditions for the markets. Therefore a dynamic description is more and more preferable.

Of course, potential applications of this dynamic theory of decision processes are obvious in the field of urban dynamics. The fundamental question why citites grow or decline under specific consideration of nested economic, social- and psychological effects has still not been answered satisfactorily. In economics we must find an acceptable way of introducing dynamics with particular emphasis being placed on the well known economic constraints. Because if equilibrium is approached as a limiting case of the dynamic equations of motion (under certain conditions) the already established and empirically tested economic results must be fulfilled in addition. The link between the micro behaviour of individual agents (consumers , firms, to mention a few) and the set of dynamic equations for appropriate macro variables is a great challenge to all of us.

The different applications and the ongoing work in this field could make us enthusiastic in thinking that *the* dynamic generalization of a large set of hitherto static approaches has been established. However, we must admit that there are different possibilities and that there are different ways required towards dynamics in socio- economic systems. In doing so researchers and policy makers will be able to find the best of the different individual approaches. The evolution of science is a complex decision process of *individual researchers* between alternatives of proceeding, discussing and deciding about the value of theories. It is my hope that the unbiased scientific decision-maker will make this contribution part of his choice set.

It goes without saying that a great deal of empirical and theoretical research is still needed for a better understanding of nested choice processes .

However, I have no doubts that we are on the way to approach Von Neumann's wish-dream and to find:

> *A dynamic theory which describes the change*
> *in terms of simpler concepts.*

Chapter 9

Appendix: The Master Equation

The *master equation* provides a fairly general mathematical method for describing the time development of any complex system (see Weidlich and Haag[1]). Before going into details of its structure, some examples will be given that illustrate the scope of its applications which ranges from physics, chemistry and biology to economics (Haken[2,3]) sociology and psychology. We shall also remark on the relation between deterministic and probabilistic descriptions of systems.

9.1 Deterministic and Probabilistic Description of Systems

Consider a system which can pass through different states in the course of time. We will assume that the number of different states I is finite, so that the index I characterizing each state is a discrete number or a set $I = \{i_1, i_2, \ldots, i_L\}$ of discrete numbers. We shall assume now that at an initial time t_o the system is in state $I_o = I(t_o)$. Two possible descriptions of the evolution over time are then feasible:

a) The information about the dynamics of the system can be *complete*. For this case the description is fully deterministic and leads to the *unique* determination of the states $I(t)$ at later times $t > t_o$. Any computer can be taken as an example for such a *deterministic system* with a finite - though very large - number of states I. Beginning with an initial state I_o set by the program, the central processor, the memory units and the peripheral devices of the computer traverse a sequence of states, $I_1, I_2, \ldots, I_N$ which are fully predetermined by the predescribed program. Finally, the unique result of the calculation represents the final state of the system.

b) The information about the system can be *incomplete.* In this case the description of the time evolution is *probabilistic* only. That means, an exact prediction of the state $I(t)$ reached by the system is not possible. Instead, the members of an ensemble of identical systems - each of them prepared in the same initial state $I(t_o)$ - will develop into *different* states $I(t)$ at time t. The best information available in this particular situation is the *probability* with which a system reaches state I at time t, given that it was prepared in state I_o at time t_o. This special probability $p(I, t|I_o, t_o)$ is denoted as *conditional probability.* The master equation will turn out to be the tool for determining this central quantity.

Comparing the deterministic with the probabilistic evolution we must state that the latter is the more general formulation, since the case of incomplete knowledge about the system comprises complete knowledge as a limit case whereas the converse is not true. The limit case of almost complete knowledge of the dynamics is revealed by the shape of the probability distribution $P(I, t)$ itself: in this case the master equation leads to an evolution in such a way that it develops one outstanding mode sharply peaked around the most likely state $I_{max}(t)$. This means that the system assumes state $I = I_{max}(t)$ with overwhelming probability at time t, whereas all the other states $I \neq I_{max}$ are highly improbable at the same time. Evidently this particular case describes a *quasi-deterministic evolution* of the system along path $I(t) \approx I_{max}(t)$.

9.2 Some General Concepts of Probability Theory

Before deriving the master equation it is useful to introduce some fundamental concepts of probability theory that especially apply to systems whose evolution are described in probabilistic terms see also the textbooks by Stratonovich,[4] Bharucha-Reid,[5] Wax,[6] Gardiner,[7] and Van Kampen.[8]

As before, it is assumed that the system can be in one of mutually exclusive different states which are characterized by a vector $\boldsymbol{i}$ consisting of one or a multiple of discrete numbers. In the course of time, transitions between states can take place. Since in general one does not know with certainty in which state the system is, the probability distribution function:

$$P(\boldsymbol{i},\ t) \tag{9.1}$$

is introduced. By definition, $P(\boldsymbol{i}, t)$ is the probability of finding the system in state $\boldsymbol{i}$ at time t. This probability has the following statistical interpretation: in an ensemble of a large number of initially equally prepared systems - so that each of them belongs to the same probability distribution - one would find systems in state $\boldsymbol{i}$ at time t with approximately the *relative frequency* $P(\boldsymbol{i}, t)$. Going to the limit case of an *infinite* ensemble its systems are found in state $\boldsymbol{i}$ with *exactly* the relative frequency $P(\boldsymbol{i}, t)$.

Since the system is with certainty in one of the states $\boldsymbol{i}$ at any time t, the probability distribution function has to satisfy condition:

$$\sum_{\boldsymbol{i}} P(\boldsymbol{i},\ t) = 1, \tag{9.2}$$

where the sum extends over all states $\boldsymbol{i}$.

Furthermore, we now introduce the most important quantity for the time evolution of the system, the *conditional probability*:

$$p(\boldsymbol{i}_2, t_2 | \boldsymbol{i}_1, t_1). \tag{9.3}$$

By definition it is the probability to find the system in state $\boldsymbol{i}_2$ at time t_2, given that it was with certainty in state $\boldsymbol{i}_1$ at time t_1. The conditional probability is fundamental for the dynamics of the system, since it describes how the probability spreads out in the time interval $(t_2 - t_1)$, given that it was concentrated on state $\boldsymbol{i}_1$ at time t_1.

The conditional probability may also depend on the *previous history* of the system, that is on states traversed *before* arriving in state I_1 at time t_1. In this general case the probability evolution process may become very complicated.

Fortunately, in many cases the so called *Markov assumption* holds, at least as a good approximation. This postulates that the evolution within time of the conditional probability $p(I_2, t_2|I_1, t_1)$ only depends on the initial state I_1 at time t_1 and state I_2 at t_2 but *not* on states of the system prior to t_1. In other words: after arriving at state I_1, the system has *lost its historic memory* and previous states do not matter in the process of further evolution. Since in many applications systems can be defined in such a way that the Markov assumption is satisfied, we shall presume it below.

We shall now subsequently draw some conclusions about the properties of the conditional probability. The following relations are a consequence of the definition:

$$p(I_2, t_1|I_1, t_1) = \delta_{I_2 I_1} \tag{9.4}$$

where

$$\delta_{I_2 I_1} = \begin{cases} 1 & \text{for } I_2 = I_1 \\ 0 & \text{for } I_2 \neq I_1 \end{cases}$$

and

$$\sum_{I_2} p(I_2, t_2|I_1, t_1) = 1. \tag{9.5}$$

Equation (9.4) follows because at time $t_2 = t_1$ the state is I_1 with probability 1. Equation (9.5) holds, since the system at any time t_2 must be in one of the states I_2 of the system.

From (9.3-5) it becomes clear that the conditional probability is the special probability distribution which evolves from the initial distribution $P(I, t_1) = \delta_{II_1}$. Furthermore, we recall the so-called *joint probability* :

$$p(I_n,t_n;\, I_{n-1},t_{n-1};\, \ldots;\, I_2,t_2;\, I_1,\, t_1). \tag{9.6}$$

By definition, this n-fold function is the joint probability to find the system in state I_1 at time t_1, in state I_2 at t_2 and ... in state I_n at t_n. From this definition, it follows that the lower order joint probabilities can be obtained from the higher ones by the following reduction formula:

$$p\,(I_3,t_3;\, I_1,t_1) = \sum_{I_2} p(I_3,t_3;\, I_2,t_2;\, I_1,t_1) \tag{9.7}$$

or in the general case, by:

$$p(I_n,t_n;\, \ldots;\, I_{k+1},t_{k+1};\, I_{k-1},t_{k-1};\, \ldots;\, I_1,t_1)$$

$$= \sum_{I_k} p(I_n,t_n;\, \ldots;\, I_{k+1},t_{k+1};\, I_k,t_k;\, I_{k-1},t_{k-1};\, \ldots;\, I_1,t_1) \tag{9.8}$$

Clearly, the summation in (9.7) over all possible states at t_2 leads to the probability of being in state I_1 at t_1 and in state I_3 at t_3 *irrespective* of the state at time t_2.

If we introduce the Markov assumption, all joint probabilities can be expressed in terms of probability (9.1) and the conditional probability (9.3). In particular, the two-fold joint probability clearly has the form:

$$p\,(I_2,t_2;\, I_1,t_1) = p\,(I_2,t_2 \,|\, I_1,t_1)\,P(I_1,t_1) \tag{9.9}$$

since the probability to find the system in state I_1 at t_1 *and* in state I_2 at t_2 is synonymous with the probability to find it in state I_1 at t_1 multiplied with the (conditional) probability to find it in state I_2 at t_2, *given that it was* in I_1 at t_1.

Generalizing this consideration and taking into account that the conditional probability does not depend on the previous history, if the Markov condition holds, we obtain:

$$p(I_n,t_n;\, I_{n-1},t_{n-1};\, \ldots;\, I_2,t_2;\, I_1,t_1)$$

$$= p(I_n,t_n|I_{n-1},t_{n-1}) \ldots p(I_2,t_2|I_1,t_1)\,P(I_1,t_1). \tag{9.10}$$

The composition formulas (9.9–10) may now be combined with the *reduction formula* (9.7) in order to derive the *Chapman-Kolmogorov equation.* Taking the sum over I_1 in (9.9) and using (9.8) yields:

$$P(I_2,t_2) = \sum_{I_1} p(I_2,t_2;\, I_1,t_1)$$

$$= \sum_{I_1} p(I_2,t_2|\, I_1,t_1)\,P(I_1,t_1)\,. \tag{9.11}$$

Equation (9.11) shows how the probability distribution $P(I_1,t_1)$ is propagated in the course of time by means of the conditional propability $p(I_2,t_2|I_1,t_1)$. Therefore, the latter is also referred to as the *propagator.* On inserting (9.9–10) in (9.7) we obtain:

$$p(I_3,t_3|\, I_1,t_1)\,P(I_1,t_1)$$

$$= \sum_{I_2} p(I_3,t_3|I_2,t_2)\,p(I_2,t_2|\, I_1,t_1)\,P(I_1,t_1). \tag{9.12}$$

Since this equation must hold for an arbitrary initial distribution $P(I_1,t_1)$ we can also conclude that:

$$p(I_3,t_3|\, I_1,t_1) \;=\; \sum_{I_2} p(I_3,t_3|I_2,t_2)\,p(I_2,t_2|\, I_1,t_1) \tag{9.13}$$

holds. Equation (9.13) is the well-known *Chapman-Kolmogorov equation.* It shows how the propagator from t_1 to t_3 can be decomposed into propagators from t_1 to t_2 and from t_2 to t_3.

9.3 The Derivation of the Master Equation

We have now seen that the propagator or in other words, the conditional probability is the crucial quantity determining the evolution with time of any probability distribution $P(I, t)$.

The master equation is nothing but a differential equation in time for the propagator: or for the probability distribution itself. It can be derived by considering equation (9.11) for times $t_1 = t$ *and* $t_2 = t + \tau$, where τ is an (infinitesimally) short time interval. Proceeding in this way, we obtain the short-time evolution:

$$P(I_2, t+\tau) = \sum_{I_1} p(I_2, t+\tau \mid I_1, t)\, P(I_1, t) \tag{9.14}$$

The short-time propagator is now being expanded in a *Taylor* series around t with respect to the variable $t_2 = t + \tau$, yielding:

$$p(I_2, t+\tau \mid I_1, t) = p(I_2, t \mid I_1, t) + \tau \left. \frac{\partial p(I_2, t_2 \mid I_1, t)}{\partial t_2} \right|_{t_2 = t} \tag{9.15}$$

$$+ \text{ higher powers in } \tau$$

Making use of (9.4) and (9.5) in (9.15) we obtain:

$$p(I_2, t \mid I_1, t) = \delta_{I_2 I_1};$$

$$\sum_{I} \left. \frac{\partial p(I, t_2 \mid I_1, t)}{\partial t_2} \right|_{t_2 = t} = 0, \tag{9.16}$$

where the sum extends over all states I of the system. Re-inserting (9.16) in (9.15) gives us:

$$p(I_2, t+\tau \mid I_1, t) = \tau\, w_t(I_2, I_1) \qquad \text{for } I_1 \neq I_2 \tag{9.17}$$

$$p(l_2, t+\tau | l_1, t) = 1 - \tau \sum_{l_1} w_t(l, l_1) \quad \text{for } l_1 = l_2 \tag{9.18}$$

where the probability transition rate:

$$w_t(l_2, l_1) = \left. \frac{\partial p(l_2, t_2 | l_1, t)}{\partial t_2} \right|_{t = t_2} \tag{9.19}$$

has been introduced for $l_1 \neq l_2$, and where the higher power terms in τ can be neglected in the limit $\tau \to 0$. Equation (9.17) states that given the system was in state l_1, the probability that it reaches state l_2 in the infinitesimally short time interval τ is proportional to that interval and to the transition rate $w_t(l_2, l_1)$ from l_1 to l_2. On the other hand, the probability to remain in the same state during the interval τ is one minus the probability transferred to all other states within the time interval (9.18).

The *master equation* is now established if (9.17, 18) is inserted in (9.14). Dividing (9.14) by τ, then after trivial rearrangements and taking the limit $\tau \to 0$ with:

$$\frac{dP(\mathbf{n}, t)}{dt} = \lim_{\tau \to \infty} \frac{P(\mathbf{n}, t+\tau) - P(\mathbf{n}, t)}{\tau}$$

the general form of the master equation is obtained:

$$\frac{dP(\mathbf{n}, t)}{dt} = \sum_{\mathbf{m}} w_t(\mathbf{n}, \mathbf{m}) P(\mathbf{m}, t) - \sum_{\mathbf{m}} w_t(\mathbf{m}, \mathbf{n}) P(\mathbf{n}, t) \tag{9.20}$$

where the sum extends over all states $\mathbf{m}$. To avoid subscripted indices we now use $\mathbf{n}$ and $\mathbf{m}$ instead if l_1 and l_2, respectively.

The master equation (9.20) is valid for any probability distribution $P(\mathbf{n}, t)$, in particular for the conditional probability $p(\mathbf{n}, t | \mathbf{m}, t')$ itself.

The master equation can be interpreted in an illustrative way. The quantity

$w_t(\boldsymbol{m}, \boldsymbol{n})$ are *transition probabilities per time unit* (or transition rates) in the following sense: $w_t(\boldsymbol{m}, \boldsymbol{n}) P(\boldsymbol{n}, t)$ is the probability transferred from state $\boldsymbol{n}$ to the state $\boldsymbol{m}$ per unit of time, which is also referred to as the *probability flux* from $\boldsymbol{n}$ to $\boldsymbol{m}$. Then (9.20) can also be read as *rate equation for probabilities.*

The change per unit of time of the probability of state $\boldsymbol{n}$ (that is the left hand side of (9.20)) is the sum of two terms with opposite effects. Firstly, there is a probability flux *from* all other states $\boldsymbol{m}$ *into* state $\boldsymbol{n}$. This is the first term of the right hand side of (9.20). Secondly, there is the probability flux *out of* state $\boldsymbol{n}$ into all other states $\boldsymbol{m}$, the second term of the right hand side of (9.20). The change per time unit of the probability $P(\boldsymbol{n}. t)$ is caused by the difference of the probability fluxes.

The master equation gives the most detailed knowledge about the evolution of a system under conditions of uncertainty or restricted information. The quantities representing this restricted information are the transition rates, namely, the transition probabilities per unit of time $w_t(\boldsymbol{m}, \boldsymbol{n})$, to reach $\boldsymbol{m}$ from $\boldsymbol{n}$. These transition rates can often be inferred from phenomenological and substantive considerations (see Chapters 3 - 8).

One solves the master equation by summing up all probability fluxes to and from each state, that is, by calculating the corresponding rise or decay in the course of time of the probabilities of each state $\boldsymbol{n}$. Equation (9.20) is a system of first-order differential equations for the evolution with time of the probability distribution $P(\boldsymbol{n}, t)$. Since the approach generates a complex system of equation for the probability distribution, both exact and approximate methods of solution are valuable. A variety of such methods, involving more or less restrictive assumptions, are available in Weidlich and Haag,[1] Haag et al,[9] and Haag.[10] For computer evaluation continued fraction solutions for the stationary probability distribution as well as the eigenvalues have proved to be a very useful tool, Haag and Hänggi,[11] Hänggi and Haag.[12]

9.4 The Stationary Solution of the Master Equation for Detailed Balance

If the transition rates $w(\mathbf{m}, \mathbf{n})$ do not explicitly depend on time t, the master equation thus describes a probability equilibration process starting with an arbitrary initial distribution $P(\mathbf{n}, t)$ and ending up with a unique final distribution $P(\mathbf{n}, \infty) = P_{st}(\mathbf{n})$. The latter is the probability distribution obeying the stationary master equation:

$$0 = \sum_{\mathbf{m}} w(\mathbf{n}, \mathbf{m}) P_{st}(\mathbf{m}) - \sum_{\mathbf{m}} w(\mathbf{m}, \mathbf{n}) P_{st}(\mathbf{n}) \quad \text{for all } \mathbf{n}. \tag{9.21}$$

In general, it is not easy to obtain a practicable form of $P_{st}(\mathbf{n})$. For a graph-theoretical solution according to the Kirchhoff theorem, see however, Weidlich.[13]

In special cases, however, the condition of *detailed balance* is fulfilled:

$$w(\mathbf{n}, \mathbf{m}) P_{st}(\mathbf{m}) = w(\mathbf{m}, \mathbf{n}) P_{st}(\mathbf{n}) \quad \text{for all } \mathbf{m}, \mathbf{n}. \tag{9.22}$$

This means that not only the global balance of all probability fluxes (9.21) hold, but that a probability flux balance between each pair of states $\mathbf{m}$. $\mathbf{n}$ holds separately.

If detailed balance (9.22) holds, the stationary solution can easily be constructed according to Haken.[14]

Take any chain $\mathcal{C}$ of states $\mathbf{n}_0 = \mathbf{0}$, $\mathbf{n}_1 = \mathbf{1}$, ..., $\mathrm{n}_{n-1} = \mathbf{n} - \mathbf{1}$, $\mathbf{n}_n = \mathbf{n}$ from a reference state $\mathbf{0}$ to an arbitrary state $\mathbf{n}$, so that *all* transition probabilities $w(\boldsymbol{\nu}, \boldsymbol{\nu} - \mathbf{1})$, $w(\boldsymbol{\nu} - \mathbf{1}, \boldsymbol{\nu})$ are nonvanishing. The repeated application of (9.22) then yields:

$$P_{st}(\mathbf{n}) = P_{st}(\mathbf{0}) \prod_{\boldsymbol{\nu}=\mathbf{0}}^{\mathbf{n}-\mathbf{1}} \frac{w(\boldsymbol{\nu} + \mathbf{1}, \boldsymbol{\nu})}{w(\boldsymbol{\nu}, \boldsymbol{\nu} + \mathbf{1})} . \tag{9.23}$$

Finally the value of $P_{st}(\mathbf{0})$ is determined by inserting (9.23) into the probability normalization condition:

$$\sum_{\mathbf{n}} P_{st}(\mathbf{n}) = 1 \qquad (9.24)$$

Condition (9.22) is not directly appropriate to check the condition of detailed balance, since it implies the stationary solution to be known beforehand. Therefore, we must derive conditions for the transition rates which are equivalent to (9.22) but do not contain the unknown stationary solution.

Since (9.22) implies (9.23) for any chain $\mathcal{C}$ and any final state $\mathbf{n}$, we can apply (9.23) also to chains of states Γ which are closed loops. As a necessary and sufficient condition of detailed balance:

$$(\Gamma) \prod_{\boldsymbol{\nu}=\mathbf{0}}^{\mathbf{n}-\mathbf{1}} \frac{w(\boldsymbol{\nu}+\mathbf{1}, \boldsymbol{\nu})}{w(\boldsymbol{\nu}, \boldsymbol{\nu}+\mathbf{1})} = 1 \qquad (9.25)$$

has to be fulfilled, for every closed loop Γ $\{\mathbf{0}, \mathbf{1}, \ldots, \mathbf{n} = \mathbf{0}\}$.

9.5 The Stationary Solution of the Master Equation of Chapter 4

In order to examine whether or not the individual movers' and suppliers' transition probabilities fulfill the condition of detailed balance:

$$w[n+l,r+k;n,r]P_{st}(n,r) = w[n,r;n+l,r+k]P_{st}(n+l,r+k)$$

$$\text{for } k,\ l = \text{neighbouring states,} \qquad (9.26)$$

we consider the closed loop:

$$\begin{array}{ccc} (n,r+1) & \leftrightarrow & (n+1,r+1) \\ \updownarrow & & \updownarrow \\ (n,r) & \leftrightarrow & (n+1,r) \end{array}$$

Writing down the detailed balance conditions at the four neighbouring points in the (n,r) space yields:

$$w[n+1,r;n,r]P_{st}(n,r) = w[n,r;n+1,r]P_{st}(n+1,r)$$

$$w[n+1,r+1;n+1,r]P_{st}(n+1,r) = w[n+1,r;n+1,r+1]P_{st}(n+1,r+1)$$

$$w[n,r+1;n+1,r+1]P_{st}(n+1,r+1) = w[n+1,r+1;n,r+1]P_{st}(n,r+1)$$

$$w[n,r;n,r+1]P_{st}(n.r+1) = w[n,r+1;n,r]P_{st}(n.r) \tag{9.27}$$

We eliminate the stationary distribution in (9.27) and obtain the result:

$$w[n+1,r;n,r]w[n+1,r+1;n+1,r]w[n,r+1;n+1,r+1]w[n,r;n,r+1]$$
$$= w[n,r;n+1,r]w[n+1,r;n+1,r+1]w[n+1,r+1;n,r+1]w[n,r+1;n,r] \tag{9.28}$$

Insertion of (4.12) in (9.28) leads to the requirement:

$$b_1 = - b_4 \tag{9.29}$$

for the trend parameters to guarantee the condition of detailed balance.

In this particular case, the *exact* stationary solution of the master equation (4.18) can be arrived at by using the chain:

$$\mathcal{C}\ ((0,0),\ (1,0),\ \ldots,\ (n,0),\ (n,1),\ldots,\ (n,r)).$$

Straightforward repeated application of (9.26) then finally leads to:

$$P_{st}(n,r) = P_{st}(0,0) \prod_{l=0}^{n-1} \frac{w[l+1,0;l,0]}{w[l,0;l+1,0]} \prod_{k=0}^{r-1} \frac{w[n,k+1;n,k]}{w[n,k;n,k+1]} \ . \tag{9.30}$$

Insertion of (4.12) into (9.30) and computing the products yields the exact solution:

$$P_{st}(n,r) = C\binom{2N}{N+n}\binom{2R}{R+r}\exp h(n,r) \tag{9.31}$$

where:

$$h(n,r) = 2b_0 n - 2b_3 r + b_2 n^2 + 2b_4 nr \tag{9.32}$$

when the detailed balance condition is met. However, the solution with detailed balance (9.31, 32) suggests that in the case without detailed balance:

$$P_{st}(n,r) = \tilde{C}\binom{2N}{N+n}\binom{2R}{R+r}\exp g(n,r), \tag{9.33}$$

where $g(n,r)$ is approximated by a polynomial of second order in n,r.

Inserting (9.33) into the stationary master equation (4.18), expanding the exponential expressions in the appropriate way, and comparing coefficients leads to the determination of $g(n,r)$ as:

$$g(n,r) = 2b_0 n - 2b_3 r + b_2 n^2 + (b_4 - b_1)\, nr. \tag{9.34}$$

Of course, the special case of detailed balance is incorporated in (9.34).

9.6 The Stationary Solution of the Master Equation of Chapter 5

It can be proved that the transition rates (5.15) satisfy the condition of detailed balance (5.20). Following the argumentation in Section 9.4 we choose a set of smallest closed chains of states which are sufficient for the proof, since arbitrarily closed chains can be composed of these smallest ones. A smallest closed chain Γ connects the following states:

$$\{\ldots,n_i,\ldots,n_j,\ldots,n_k,\ldots\} = \mathbf{0} \to \{\ldots,(n_i-1),\ldots,(n_j+1),\ldots,n_k,\ldots\} = \mathbf{1}$$

$$\to \{\ldots,(n_i-1),\ldots,n_j,\ldots,(n_k+1),\ldots\} = \mathbf{2} \to \{\ldots,n_i,\ldots,n_j,\ldots,n_k,\ldots\} = \mathbf{0}, \tag{9.35}$$

and corresponds to the ring migration of a member of the population between areas $i \to j \to k \to i$. Only one term of the right hand side of (5.15) contributes to each of the transitions (9.35). Hence, formula (9.25) reduces to (to be read from right to left):

$$\frac{w(\mathbf{0},\mathbf{2})}{w(\mathbf{2},\mathbf{0})} \cdot \frac{w(\mathbf{2},\mathbf{1})}{w(\mathbf{1},\mathbf{2})} \cdot \frac{w(\mathbf{1},\mathbf{0})}{w(\mathbf{0},\mathbf{1})}$$

$$= \frac{w_{ik}(n_i-1,n_k+1)}{w_{ki}(n_k,n_i)} \cdot \frac{w_{kj}(n_k,n_j+1)}{w_{jk}(n_j,n_k+1)} \cdot \frac{w_{ji}(n_j,n_i)}{w_{ij}(n_i-1,n_j+1)} = 1, \tag{9.36}$$

which can easily be checked. This completes the proof that detailed balance is fulfilled.

The most important consequence of detailed balance is that the stationary solution of the master equation can be constructed using equation (9.23). For this aim we shall consider a chain of states:

$$\{N,0,\ldots,0\} \to \{N-1,1,0,\ldots,0\} \to \{N-2,2,0,\ldots,0\} \to \{N-n_2,n_2,0,\ldots,0\}$$
$$\to \{N-n_2-1,n_2 1,0,\ldots,0\} \to \{N-n_2-n_3,n_2,n_3,0,\ldots,0\} \to \{n_1,n_2,n_3,\ldots,n_L\}$$
$$\tag{9.37}$$

where $N = \sum_i^L n_i$. We start from the reference state $\{N,0,\ldots,0\}$ and end up with the general state $\{n_1,n_2,\ldots,n_L\}$. They are connected by nonvanishing transition rates. Hence, we can use this chain to construct $P_{st}(n_1,n_2,\ldots,n_L)$ from $P_{st}(N,0,\ldots,0)$ according to (9.23). By inserting into (9.23) the transition rates (9.35), which connect these states, we obtain for instance, the intermediate result:

$$P_{st}(N-n_2, n_2, 0, \ldots, 0)$$

$$= \frac{N(N-1)\cdots(N-n_2+1)}{n_2!} \cdot \exp\left\{2\sum_{m=1}^{n_2} u_2(m) - 2\sum_{m=N-n_2+1}^{N} u_1(m)\right\} \cdot P_{st}(N,0,\ldots,0) \tag{9.38}$$

Continuing the procedure along chain (9.37), we finally obtain result (5.21).

9.7 The Embedding of Random Utility Theory

Choice problems are concerned with a population of individuals who have to select an alternative among a set of alternatives. One of the most popular individual choice models is the multinomial logit model. A great advantage of this model is that it gives explicit expressions of the choice probabilities. The multinomial logit model shortly presented here, however, is limited in its usefulness because it does not account for the effects of the time factor and the social interactions between individuals (de Palma[15]). In this section we want to show the relation of the multinomial logit model to our dynamic framework of decision processes.

9.7.1 The Multinomial Logit Model

In the random utility theory, we consider an individual k, who has to choose an alternative A_i , among a set of J mutually exclusive alternatives A_j, j = 1, 2, ..., J . The individual chooses the option which is considered most desirable to it, given the attributes z_j^k of each alternative A_j as seen by the individual k and his personal tastes s^k. More precisely, an individual k measures the desirability of each alternative A_j by a utility function u_j^k which can be expressed as the sum of two components, that is :

$$u_j^k = v_j^k(z_j^k, s^k) + \mu^k \varepsilon_j^k, \qquad \text{for } j = 1, 2, \ldots, J \tag{9.39}$$

where the term $v_j^k(z_j^k, s^k)$ is non-stochastic and contains all the attributes of A_j and characteristics which are observed by the individual. Therefore, this is the deterministic part of the utility function. The term ε_j^k is stochastic (the random part) and reflects the effect of all the unobserved variables (the uncertainty). The coefficient μ^k is a positive coupling constant that measures the importance of the error terms ε_j^k A clear interpretation of this parameter is given by de Palma and Lefèvre.[16] It is then assumed that the individual chooses the alternative A_j , if this is the choice which maximizes his utility. Therefore, the probability p_j that an individual having a utility function u_j^k selects A_j is given by:

$$p_j^k = \text{Prob}\left[v_j^k(z_j^k, s^k) + \mu^k\varepsilon_j^k \geq v_i^k(z_i^k, s^k) + \mu^k\varepsilon_i^k \; \forall i \neq j \right],$$

$$j = 1, 2, \ldots, J \qquad (9.40)$$

The classical multinomial logit model corresponds to the case where the random utility parts ε_j^k are assumed to be independent *Gumbel* distributed. At this point the assumption of stationarity comes in. The probabilities p_j^k can then be written very simply as (Domencich and McFadden[17]):

$$p_j^k = \frac{\exp(v_j^k/\mu^k)}{\sum_{i=1}^{J} \exp(v_i^k/\mu^k)}, \qquad \text{for } j = 1, 2, \ldots, J \qquad (9.41)$$

This model possesses some other very interesting properties which contribute to its wide use in economics, geography, biology as well as psychology. For further applications see also Leonardi.[18]

9.7.2 The Multinomial Logit Model as Limiting Case of our Dynamic Theory

Let us consider a group of N individuals. Each individual has to select one alternative A_j out of a set of $j = 1, 2, \dots, J$ alternatives. We, however, are now interested in the *dynamics* of the decision configuration:

$$\boldsymbol{n} = \{n_1, n_2, \dots, n_J\}, \tag{9.42}$$

where n_j is the number of individuals who have decided for alternative A_j. Changes in the decision configuration are assumed to be caused by differences in *"dynamic"* utilities u_j. The u_j measures the desirability of alternative j. The dynamic utilities enter the transition rates $p_{ji}(\boldsymbol{n})$ that an individual who originally preferred alternative i now makes a transition to alternative j.

In our dynamical context, the transition rates have the following form in terms of dynamic utilities:

$$p_{ji}(\boldsymbol{n}) = \nu \exp[u_j(\boldsymbol{n}) - u_i(\boldsymbol{n})] , \tag{9.43}$$

where ν is a flexibility parameter.

Note here, that the dynamic utilities in general can also depend on the decision configuration (9.42) and therefore can include interaction effects among individuals.

The total transition rate then is given by:

$$w_{ji} = n_i \, p_{ji}(\boldsymbol{n}) = n_i \nu \exp[u_j(\boldsymbol{n}) - u_i(\boldsymbol{n})] \tag{9.44}$$

Since the different individuals have to choose one or the other alternative, the dynamics of decision processes in general are mathematically similar to migration processes (compare Chapter 5).

The corresponding master equation reads:

$$\frac{dP(\boldsymbol{n},t)}{dt} = \sum_{i,j}^{J} w_{ji}(\boldsymbol{n}^{(ij)})P(\boldsymbol{n}^{(ij)},t) - \sum_{i,j}^{J} w_{ij}(\boldsymbol{n})P(\boldsymbol{n},t) \tag{9.45}$$

where $P(\boldsymbol{n}, t)$ is the probability to find a certain decision configuration realized at time t (notations according to Chapter 5). In contrast to the multinomial logit model, the master equation contains the full stochastics and dynamics of the decision process, for instance starting from any initial probability distribution $P(\boldsymbol{n}, 0)$.

The exact stationary solution $P_{st}(\boldsymbol{n})$ of the master equation (9.45), assuming the transition rates (9.44) is reached for $t \to \infty$. Its explicit form was found in Section 5.2.4.

This means that not only the most probable evolution and final state of the decision behaviour of the considered group of individuals is obtained but also its full stochastics.

The most probable stationary decision configuration $\hat{\boldsymbol{n}} = \{\hat{n}_1, \hat{n}_2, \ldots, \hat{n}_J\}$ is given by (see Section 5.2.4):

$$\hat{n}_j = \frac{N \exp[2u_j(\hat{\boldsymbol{n}})]}{\sum_{i=1}^{J} \exp[2u_i(\hat{\boldsymbol{n}})]} \qquad \text{for } j = 1, 2, \ldots, J \tag{9.46}$$

where n_j is the most probable number of individuals who have decided for alternative A_j.

Thus the quantity:

$$\hat{p}_j = \frac{\hat{n}_j}{N} = \frac{\exp[2u_j(\hat{\boldsymbol{n}})]}{\sum_{i=1}^{J} \exp[2u_i(\hat{\boldsymbol{n}})]} \qquad \text{for } j = 1, 2, \ldots, J \tag{9.47}$$

is equivalent to the probability that any one individual selects alternative A_j.

Comparing the stationary solution (9.47) of our dynamic theory of choice processes with the outcome of the multinomial logit model one has to identify:

$$u_j(\hat{\boldsymbol{n}}) = \frac{1}{2} v_j^k / \mu^k . \tag{9.48}$$

Thus both "*utilities*" coincide up to an ordinary rescaling, if the same utility function (9.48) can be assigned to all individuals k of the decision configuration.

The coincidence of the stationary formula (9.47) with the formula (9.41) of the multinomial logit model under appropriate rescaling (9.48) of the utility concepts has the meaning, that the multinomial logit model describes the limiting case for $t \to \infty$ of our dynamic description in the special case of *noninteracting individuals*, where u_j does not depend on $\boldsymbol{n}$.

9.8 The Construction of Configurational Transition Rates via Panel Data

Panel data provide the information on a sample of decision-makers (agents) at different points in time. Therefore, statements can be made about the behavioural response at the microlevel. As already mentioned dynamic models of discrete choice have to be based on panel data. In this section we shall briefly outline how the constitutive elements of the master equation, namely the configurational transition rates can be obtained from such an analysis.

The configurational transition rates are defined via the conditional probabilities according to (2.15), (9.19):

$$w_t(\boldsymbol{n} + \boldsymbol{k}, \boldsymbol{n}) = \left. \frac{\partial p(\boldsymbol{n} + \boldsymbol{k}, t' \,|\boldsymbol{n}, t)}{\partial t'} \right|_{t' = t} \tag{9.49}$$

By definition, the configurational conditional probability $p(\boldsymbol{n}', t+\tau\,|\boldsymbol{n}, t)$ is the probability to find configuration $\boldsymbol{n}'$ at time $t+\tau$, given that the configuration $\boldsymbol{n}$ was realized at time t. On the other hand, the *individual* conditional probability $p^{(l)}(i, t+\tau|j, t)$ is the probability to find the *individual* decision-maker l in state i of behaviour at time $t+\tau$, given that he was in state j at time t. The subtle combinatorical relationship between both kinds of conditional probabilities can be solved if the time interval τ is sufficiently small (Weidlich and Haag[19]) and the individual decision acts can be treated as statistically independent. However, the decisions of the individuals still are allowed to depend on the realized decision configuration $\boldsymbol{n}$. Groups of individuals (e.g., households, couples) can be treated in the same manner. Even in the case of small correlations between the decision behaviour of different agents the linear relationship (9.51) seems to be justified.

The configurational transition rate can now be decomposed (see (2.25) into several terms, where the different terms are responsible for different effects (compare Section 2.3):

$$w_t(\boldsymbol{n}+\boldsymbol{k}, \boldsymbol{n}) = \sum_{j,i}\sum_{\alpha} w_{ji}^{\alpha}(\boldsymbol{n}+\boldsymbol{k}, \boldsymbol{n}) + \sum_{\alpha,\beta}\sum_{i} w_{i}^{\beta\alpha}(\boldsymbol{n}+\boldsymbol{k}, \boldsymbol{n}) + \sum_{\alpha,i}\{w_{i+}^{\alpha}(\boldsymbol{n}+\boldsymbol{k}, \boldsymbol{n}) + w_{i-}^{\alpha}(\boldsymbol{n}+\boldsymbol{k}, \boldsymbol{n})\}, \tag{9.50}$$

In the following we shall restrict ourselves to the important case that an agent changes his attitude (compare Section 2.3.3). As already mentioned this kind of process is formally equivalent to migratory processes and therefore crucial in the context of this book. In (2.19) we assumed that the configurational transition rate:

$$w_{ij}^{\alpha}(\boldsymbol{n}+\boldsymbol{k}, \boldsymbol{n}) = n_{\alpha j}\, p_{ij}^{\alpha}(\boldsymbol{n}, \boldsymbol{x}) \tag{9.51}$$

$$\text{for} \quad \boldsymbol{k} = \{0, \ldots, 1_{\alpha i}, \ldots, 0, \ldots, -1_{\alpha j}, \ldots, 0, \ldots\},$$

is proportional to the number of individuals of subpopulation α having adopted state j of behaviour (living in region j) multiplied by an "individual" transition rate $p_{ij}^{\alpha}(\boldsymbol{n}, \boldsymbol{x})$. This relation (9.51) can be derived via the above mentioned assumptions. The "individual" transition rates may depend on the decision configuration itself. Hence, synergetic (selfreinforcing) mechanisms are included in this general concept.

However, the p_{ij}^{α} are here *not* individual transition rates in the sense, that different transition rates must be attributed to different individuals. Instead we suppose that the different agents of the system (the individuals) belonging to the same ensemble α *operate* according to the same probability transition rate p_{ij}^{α} depicting the probabilistic decision behaviour of each member of the ensemble α.

If, however, panel data are available on the decision behaviour of a group of agents (micro-level) we are able to substitute the assumption (9.51) by direct computation of the configurational transition rates via:

$$w_{ij}^{\alpha}(\boldsymbol{n}+\boldsymbol{k},\boldsymbol{n}) = \sum_{l \in \mathcal{Q}_j} p_{ij}^{\alpha(l)}(\boldsymbol{n},\boldsymbol{x}) \tag{9.52}$$

where we have to sum up over all individual contributions $p_{ij}^{\alpha(l)}$, of all agents $l \in \mathcal{Q}_j$ belonging to subpopulation α, and changing from state j to state i of behaviour per time unit τ. The $p_{ij}^{\alpha(l)}$ can be obtained from the panel data using the definitions (9.49),(2.19),(2.25).

It is worth mentioning that via a comparison of (9.52) with (2.19), (2.32) the functional dependence of the $p_{ij}^{\alpha} = p_{ij}^{\alpha}(\boldsymbol{n},\boldsymbol{x})$ on socio-economic characteristics can be tested.

References

Chapter 1

[1] J. von Neumann and M. Morgenstern, *Theory of Games and Economic Behaviour.*

[2] J.-P. Aubin, D. Saari, K. Sigmund, *Dynamics of Macrosystems*, (Springer, Berlin, Heidelberg, New York, 1985).

[3] A.de Palma, *Individual Decision-Making in Dynamic Collective Systems, Part 1 and Part 2*, Journal of Mathematical Sociology, forthcoming (1988).

[4] H.Haken, *Synergetics, an Introduction*, 2nd. ed., Springer Series Synergetics, Vol.1 (Springer, Berlin, Heidelberg, New York, 1977).

[5] H.Haken, *Cooperative Phenomena in Systems far from Equilibrium and in Nonphysical Systems*, Rev. Mod. Phys. **47**, 67 (1975).

[6] A.Pacault,C.Vidal(eds.), *Synergetics far from Equilibrium*, Springer Ser.Synergetics, Vol.3 (Springer, Berlin, Heidelberg, New York, 1978).

[7] W.Güttinger, H.Eikemeier (eds.), *Structural Stability in Physics*, Springer Ser. Synergetics, Vol.4 (Springer, Berlin, Heidelberg, New York, 1978).

[8] L.A.Blumenfeld, *Problems of Biological Physics*, Springer Ser. Synergetics, Vol. 7 (Springer, Berlin, Heidelberg, New York, 1979) and L.A. Blumenfeld: *Physics of Bioenergetic Processes*, Springer Ser. Synergetics, Vol.16 (Springer, Berlin, Heidelberg, New York, 1981).

[9] W.Weidlich, *The use of statistical mathods in sociology*, Collect. Phenomena, **1**, 51 (1972).

[10] W.Weidlich, *The Statistical Description of Polarization Phenomena in Society*, Br. J. math. statist. Psychol., **24**, 251-266 (1971).

[11] W.Weidlich, G.Haag, *Concepts and Models of a Quantitative Sociology, The Dynamics of Interacting Populations*, Springer Ser. Synergetics,

Vol.14 (Springer, Berlin, Heidelberg, New York, 1983).

[12] H. Haken, *Advanced Synergetics, Instability Hierarchies on Self-Organizing Systems and Devices*, Springer Series in Synergetics, Vol.20 (Springer, Berlin, Heidelberg, New York, 1983).

[13] H.P.Schwefel, *Numerical Optimization of Computer Models*, (Wiley, Chichester, 1981).

Chapter 2

[1] R.G.Golledge and H.Timmermans (eds.), *Behavioural Modelling in Geography and Planning*, (Croom Helm, London, New York and Sydney, 1988).

[2] M.Ben-Akiva and S.R.Lerman, *Discrete Choice Analysis: Theory and Application to Predict Travel Demand*, (MIT Press, Cambridge, MA and London, 1985).

[3] E. Bahrenberg, M.M.Fischer and P.Nijkamp (eds.), *Recent Developments in Spatial Analysis: Methodology, Measurement, Models*, (Gower, Aldershot, 1984).

[4] D.E.Pitfield (ed.), *Discrete Choice Models in Regional Science*, (Pion, London, 1984).

[5] L.W.Johnson and D.A.Hensher, *Application of Multinomial Probit to a Two-Period Panel Data Set*, Transportation Research, **16A**, 457-464 (1982).

[6] M.M.Fischer, G.Haag, M.Sonis, W.Weidlich, *Account of Different Views in Dynamic Choice Processes*, Paper presented at the IGU- Working Group on Mathematical Models, Canberra, August 16-19, (1988).

[7] D.A.Hensher, *Model Specification for a Dynamic Discrete Continuous Choice Automobile Demand System*, in: R.G.Golledge and H.Timmermans (eds.): *Behavioural Modelling in Geography and Planning*, (Croom Helm, London, New York and Sydney, 1988).

[8] J.S.Coleman, *Longitudinal Data Analysis*, (Basic Books Inc. New York, 1981).

[9] N.B.Tuma and M.G.Hannen, *Social Dynamics Models and Methods* (Aca-

demic Press, Orlando et.al., 1984).

[10] D.A.Hensher and N.Wrigley, *Statistical Modelling of Discrete Choice with Panel Data*, Working Paper, **16**, School of Economic and Financial Studies, Macquarie University, Australia (1984).

[11] N.Wrigley, *Quantitative Methods: The Era of Longitudinal Data Analysis*, Progress in Human Geography, **10**, 84-102 (1986).

[12] J.J.Heckman, *Statistical Models for Discrete Panel Data,* in: C.F. Manski and D.McFadden (eds.): *Structural Analysis of Discrete Panel Data with Economic Applications*, (MIT Press, Cambridge, MA and London, 1981).

[13] T. Domencich, D. McFadden, *Urban Travel Demand: A Behavioural Analysis*, (North Holland, Amsterdam, 1975).

[14] W.Weidlich, *The Statistical Description of Polarization Phenomena in Society*, Br. J. math. statist. Psychol., **24**, 251-266 (1971).

[15] W.Weidlich, G.Haag, *Concepts and Models of a Quantitative Sociology, The Dynamics of Interacting Populations*, Springer Ser. Synergetics, Vol.14 (Springer, Berlin, Heidelberg, New York, 1983).

[16] T.R.Smith, *Transition Probabilities and Behaviour in Master Equation Descriptions of Population Movements*, in: D.A.Griffith and R. MacKinnon (eds.): *Dynamic Spatial Models* (Sijthoff and Noordhoff, Alphen aan den Rijn, 1981).

[17] G.Haag, W. Weidlich, *A Stochastic Theory of Interregional Migration*, Geographical Analysis, **16**, 331-357 (1984).

[18] G. Haag, W. Weidlich, *A Dynamic Migration Theory and its Evaluation for Concrete Systems*, Regional Science and Urban Economics, **16**, 57-80 (1986).

[19] G.Haag and W.Weidlich, *A Non-Linear Dynamic Model for the Migration of Human Populations*, in: D.A.Griffith and A.Lea (eds.): *Evolving Geographical Structures* (Martinus Nijhoff, The Hague, Boston and Lancaster, 1983).

[20] G. Haag, D.S.Dendrinos, *Toward a Stochastic Theory of Location: A Nonlinear Migration Process,* Geographical Analysis, **15**, 269-286 (1983).

[21] D.S.Dendrinos, G.Haag, *Toward a Stochastic Theory of Location: Empirical Evidence,* Geographical Analysis, **16**, 287-300 (1984).

[22] G.Leonardi, *Transient and Asymptotic Behaviour of a Random Utility Based Stochasitic Search Process in Continuous Space and Time*, Collaborative Paper, WP-83- , Laxenburg (1983).

[23] M.Wegener, *Linking Spatial Choice Models*, Paper presented at the Workshop on Spatial Choice Models in Housing, Transportation, and Land Use, IIASA (1982).

[24] R.D. Luce, *Individual Choice Behaviour*, (Wiley, New York, 1959)

[25] W.Weidlich, G. Haag(eds.), *Interregional Migration, Dynamic Theory and Comparative Analysis* (Springer, Berlin, Heidelberg, New York 1988).

[26] J.-M. Huriot, J. Thisse, *Distance in Spatial Analysis*, Institute de Mathématiques Economiques, LATEC L.A., CNRS **342**, Universite' de Dijon (1984).

[27] J. H. Kuiper, *Distance Distributions in European Countries*, paper presented at the 5th European Colloquium of Quantitative and Theoretical Geography, Bardonecchia (1987).

[28] N.S. Goel, N. Richter-Dyn, *Stochastic Models in Biology* (Academic, New York 1974).

[29] W.M. Getz, *Stochastic Equivalents of the L inear and Volterra- Lotka Systems of Equations*, Mathematical Biosciences, **29**, 235 (1976).

[30] N.S. Goel, S.C. Maitra, E.W. Montroll, *On the Volterra and other Nonlinear Models of Interacting Populations*, Review of Modern Physics, **43**, 231 (1971).

[31] E.C.Pielou, *An Introduction to Mathematical Ecology* (Wiley-Interscience, New York 1969).

[32] D. Ludwig, *Stochastic Population Theories*, Lecture Notes Biomathematics, **3** (Springer, Berlin. Heidelberg, New York 1974).

[33] G. Eilenberger, H. Müller-Krumbhaar, *Nichtlineare Dynamic in kondensierter Materie*, Proceedings der Ferienschule, Kernforschungsanlage, Jülich (1983).

[34] A. Sen, *Maximum Likelihood estimation of gravity model parameters*, Journal of Regional Science, (forthcoming).

[35] A. Sen, R.K. Pruthi, *Least squares calibration of gravity model parameters when intrazonal flows are unknown*, Environment and Planning A, **15**, 1545-1550 (1983).

[36]H.P. Schwefel, *Numerical Optimization of Computer Models* (Wiley, Chichester, 1981).

[37]H.P. Schwefel, F. Drepper, R. Heckler, *Combining Estimation, Simulation and Optimization in Computer-Aided Energy Planning*, Angewandte Systemanalyse, **2**, 69-79 (1981).

[38]F. Drepper, U. Hermes, *IRECA - Interface for Regression and Correlation Analysis - User's Manual*, Internal Report, Nuclear Research Center (KFA), Jülich (1979).

[39]F. Drepper et al., *DAIMOS - Data Interface for Modular Dynamic Simulation - User's Manual*, Internal Report, Nuclear Research Center (KFA), Jülich (1981).

[40]R. Heckler, *OASIS - Optimization and Simulation Integrating Systems*, Internal Report, Nuclear Research Center (KFA), Jülich (1979).

[41] A. Sen, *Estimation of Parameters for the Poisson Gravity Model*, Paper presented at the Research and Development, Industrial Change and Economic Policy (RICE) Symposium, Karlstad, 22- 26 June, 1987.

[42]H.P. Schwefel, *Collective Phenomena in Evolutionary Systems*, Paper presented at *Problems of Constancy and Change*, 31st Annual Meeting of the International Society for General Systems Research, Budapest (1987).

[43]R. Reiner, M. Munz, *Ranking Regression Analysis of Spatio- Temporal Variables*, Theory and Decision, to be published.

Chapter 3

[1] Y.Y. Papageorgiou, *On Sudden Urban Growth*, Environment and Planning A, **12**, 1035-1050 (1980).

[2]W.C. Wheaton, *A Comparative Static Analysis of Urban Spatial Structure*, Journal of Economic Theory, **9**, 223-237 (1974).

[3] E. Casetti, *Equilibrium Population Partitions between Urban and Agricultural Occupations*, Geographical Analysis, **12**, 47-54 (1970).

[4]P.Ehrenfest, T.Ehrenfest, Enc. Math. Wiss. (1911).

[5] G. Haag, *A Stochastic Theory on Sudden Urban Growth*, Konzepte SFB 230, Heft 25, 255-271 (1986).

[6] M. Abramowitz, I.A. Stegun, *Handbook of Mathematical Functions*, (U.S. Department of Commerce, N.B.S. Appl. Math. **55**, 1964).

Chapter 4

[1] G. Haag, D.S.Dendrinos, *Toward a Stochastic Theory of Location: A Nonlinear Migration Process*, Geographical Analysis, **15**, 269-286 (1983).

[2] D.S.Dendrinos, G.Haag, *Toward a Stochastic Theory of Location: Empirical Evidence*, Geographical Analysis, **16**, 287-300 (1984).

[3] D.S.Dendrinos, *Urban Evolution, Studies in the Mathematical Ecology of Cities* (Oxford University Press, 1985).

[4] J.H. von Thünen, *Der isolierte Staat in Beziehung auf Landwirtschaft und Nationalökonomie*, Hamburg (1826).

[5] M.Beckmann, *On the Distribution of Rent and Residential Density in Cities*, paper presented at the Inter-Departmental Seminar on Mathematical Applications in the Social Sciences, Yale University

[6] W. Alonso, *Location and Land Use* (Cambridge, Mass.: Harvard University Press, 1964).

[7] E. Mills, *Studies in the Structure of the Urban Economy* (Baltimore, Md.: Johns Hopkins University Press, 1972).

[8] T. Miyao, *Dynamic Analysis of the Urban Economies* (New York: Academic Press, 1981).

[9] J. Pack, *Urban Models: Diffusion and Policy Application*: Monograph Series No.7, Regional Science Research Institut, University of Pennsylvania, 1978.

[10] A.G. Wilson, *Entropy in Urban and Regional Modelling*, (London: Pion, 1970).

[11] A.G. Wilson, *Catastrophe Theory and Bifurcation. Applications to Urban and Regional Systems* (London, Croom Helm, 1981).

[12] P. Allen, J.L.Deneubourg, M.Sanglier, A.DePalma, *The Dynamics of Urban*

Evolution, Final Report to the US Department of Transportation, Washington, DC., 1978.

[13] P. Allen, M.Sanglier, *A Dynamic Model of a Central Place System*, Geographical Analysis, **13**, 149-164 (1981).

[14] J. Forrester, *Urban Dynamics* (Cambridge, Mass.: MIT Press 1971).

Chapter 5

[1] D. Courgeau, *Interaction between spatial mobility, family and carer life-cycle: A French survey*, European Sociological Review, **1**, 139-162 (1985).

[2] D.Courgeau, *Migrants and Migrations*, Population Selected Papers (Editions de l'I.N.E.D., Paris, 1979).

[3] N.Keyfitz, *Introduction to the Mathematics of Population* (Addison Wesley, 1968).

[4] H. Hotelling, *A Mathematical Theory of Migration*, republished in Environment and Planning A, 10 (1978).

[5] T. Puu, *A Simplified Model of Spatio-Temporal Population Dynamics*, Umea Economic Studies, 139 (1983).

[6] Å.E. Andersson, G. Ferraro, *Accessibility and Density Distributions in Metropolitan Areas: Theory and Empirical Studies*, Papers of the Regional Science Association, 52 (1983).

[7] Å.E. Andersson, A. Karlquist, *Population and Capital in Geographical Space*, J. Los, M. Los (eds.), Computing Equilibria: How and Why (North Holland, 1976).

[8] Å.E. Andersson,D. Philipov, *Economic Models of Migration, Regional Development Modeling: Theory and Practice*, M. Albegov, A.E. Andersson, F. Snickars (eds.) (North Holland, 1982).

[9] M.J. Beckmann, *Spatial Equilibrium in the Dispersed City,* Mathematical Land Use Theory, Y.Y. Papageorgiou (ed.) (Lexington Books, 1976).

[10] G. Leonardi, J. Casti, *Agglomeration Tendencies in the Distribution of Populations*, Regional Science and Urban Economics, **16**, (1986).

[11] W.A. Clark, T.R. Smith, *Housing Market Search Behaviour and Expected Utility Theory*, Environment and Planning A, **14**, 681- 698 (1982).

[12] W.A. Clark, J.E. Burt, *The Impact of Workplace on Residential Location*, Annals, Association of American Geographers, **70**, 59-67 (1980).

[13] D.Pumain, *Evolving Structure of the French Urban System*, Urban geography, **5**, 308-325 (1984).

[14] L. Curry, *Division of Labour from Geographical Competition*, Annals, Association of American Geographers, **71**, 133-165 (1981).

[15] D.A.Griffith, *Dynamic Characteristics of Spatial Economic Systems*, Economic Geography, **58**, 178-196 (1982).

[16] R. MacKinnon, *Dynamic Programming and Geographical Systems*, Economic Geography, **46**, 350-366 (1970).

[17] M.Sonis, *Flows, Hierarchies, Potentials*, Environment and Planning A, **13**, 413-420 (1981).

[18] D.A. Griffith, A.C. Lea (eds.), *Evolving Geographical Structures,* (The Hague: Martinus Nijhoff, 1983).

[19] G.Haag,W. Weidlich, *A Stochastic Theory of Interregional Migration*, Geographical Analysis, **16**, 331-357 (1984).

[20] G. Haag, W. Weidlich, *A Dynamic Migration Theory and its Evaluation for Concrete Systems*, Regional Science and Urban Economics, **16**, 57-80 (1986).

[21] W. Weidlich, G. Haag (eds.), *Interregional Migration, Dynamic Theory and Comparative Analysis* (Springer, Berlin, Heidelberg, New York, 1988).

[22] D. Pumain, *La Dynamique des Villes* (Paris, Econometrica, 1982).

[23] R. Reiner, M. Munz, *Ranking Regression Analysis of Spatio- Temporal Variables*, Theory and Decision, to be published.

[24] R.L. Stratonovich, *Topics in the Theory of Random Noice*, Vol.1 and Vol.2 (Gordon and Breach, New York 1963 and 1967).

[25] W. Weidlich, G. Haag, *A Dynamic Phase Transition Model for Spatial Agglomeration Processes*, forthcoming.

[26] H. Birg, *Analyse und Prognose der Bevölkerungsentwicklung in der Bundesrepublik Deutschland und in ihren Regionen bis zum Jahr 1990*, Deut-

sches Institut für Wirtschaftsforschung, **35** (1975).

[27] H. Birg; J. Huinink, H. Koch, H. Vorholt, *Kohortenanalytische Darstellung der Geburtenentwicklung in der Bundesrepublik Deutschland*, IBS-Materialien, **10**, Universität Bielefeld (1984).

[28] H. Birg, D. Filip, K. Hilge, *Verflechtungsanalyse der Bevölkerungsmobilität zwischen den Bundesländern von 1950 - 1980*, IBS-Materialien, **8**, Universität Bielefeld (1983).

[29] R. Koch, H.-P. Gatzweiler, *Migration and Settlement: 9. Federal Republik of Germany*, IIASA, RR - 80 - 37, Laxenburg (1980).

Chapter 6

[1] H.Haken: *Advanced Synergetics*, Springer Ser. Synergetics, Vol. 20 (Springer, Berlin, Heidelberg, New York, 1983).

[2] D.S.Dendrinos: *Urban Evolution, Studies in the Mathematical Ecology of Cities* (Oxford University Press, 1985).

[3] E. Mosekilde, S. Rasmussen, H. Joergensen, F. Jaller, C. Jensen, *Chaotic Behaviour in a Simple Model of Urban Migration*, preprint, The Technical University of Denmark (1985).

[4] W. Weidlich, G. Haag, *Migration of Mixed Population in a Town*, Collective Phenomena, **3**, 89 (1980).

[5] G. Haag, W. Weidlich, *A Stochastic Theory of Interregional Migration*, Geographical Analysis, **16**, 331-357 (1984).

[6] R. Reiner, M. Munz, G. Haag, W. Weidlich, *Chaotic Evolution of Migratory Systems*, Sistemi Urbani, **2/3**, 285-308 (1986).

[7] H.G.E. Hentschel, I. Procaccia, *The Infinite Number of Generalized Dimensions of Fractals and Strange Attractors*, Physica 8D, 435-444 (1983).

[8] G. Benettin, L. Galgani, J.-M. Strelcyn, *Kolmogorov Entropy and Numerical Experiments*, Phys. Rev. A, **14**, 2338 (1976).

[9] J.L. Kaplan, J.A. Yorke, *Functional Differential Equations and Approximations of Fixed Points*, ed. by H.-O. Peitgen and H.-O. Walther, Lecture

Notes in Mathematics, 730 (1979).

[10] P. Grassberger, I. Procaccia, *Measuring the Strangeness of Strange Attractors*, Physica D, **9**, 189 (1983).

[11] J.D. Farmer, E. Ott, J.A. Yorke, Physica **4D**, 366 (1982).

Chapter 7

[1] C.S. Bertuglia, G. Leonardi, A.G. Wilson (eds.), *Urban Systems, Designs for an Integrated Dynamic Model*, forthcoming (1988).

[2] J.R. Roy, *Estimation of Singly-Constrained Nested Spatial Interaction Models*, Environment and Planning B, **10**, 269-274 (1983)

[3] J.R. Roy, *On Forecasting Choice among Dependent Spatial Alternatives*, Environment and Planning B, **12**, 479-492 (1985)

[4] G. Leonardi, *Optimal Facility Location by Accessibility Maximizing*, Environment and Planning A, **10**, 1287-1305 (1978)

[5] G. Leonardi, A Unifying Framework for Public Facility Location Problems, Environment and Planning A, **13**, 1001-1028 (1981)

[6] L.D.Mayhew and G.Leonardi, *Equity, Efficiency and Accessibility in Urban and Regional Health-Care Systems*, Environment and Planning A, **11**, 177-192 (1979)

[7] J.R.Roy and P.F.Lesse, *Planning Models for Non-Cooperative Situations: A Two-Player Game Approach*, Regional Science and Urban Economics, **13**, 205-221 (1983)

[8] J.R.Roy, B.Johansson, G.Leonardi, *Some Spatial Equilibria in Facility Investment under Uncertain Demand*, Paper of the Regional Science Association, **56**, 215-228 (1985)

[9] T.R.Lakshmanan and W.G.Hansen, *A Retail Market Potential Model*, Journal of the American Institute of Planners, **31**, 134-143 (1965)

[10] A.G. Wilson, *A family of Spatial Interaction Models and Associated Developments*, Environment and Planning, **3**, 1-32 (1971).

[11] B. Harris, A.G. Wilson, *Equilibrium Values and Dynamics of Attractiveness Terms in Production-Constrained Spatial- Interaction Models*,

Environment and Planning A, **10**, 371-388 (1978).

[12] F.M. Allen, M. Sanglier, *Dynamic Models and Urban Growth*, Journal of Social and Biological Structures, **1**, 265-280, and **2**, 269-278 (1978).

[13] I.S. Lowry, *A Model of Metropolies*, (Rand Corporation, Santa Monica, 1964).

[14] D.M. Hill, *A Growth Allocation Model for the Boston Region*, Journal of the American Institute of Planners, **31**, 111-120 (1965).

[15] J. Forrester, *Urban Dynamics*, (MIT Press, Cambridge, 1968).

[16] E.S. Mills, *Studies in the Structure of Urban Economies*, (John Hopkins Press, 1972).

[17] D. Pumain, *La Dynamique des Villes* (Paris, Econometrica, 1982).

[18] D. Pumain, Th. Saint-Julien, L. Sanders, *Dynamics of Spatial Structure in French Urban Agglomerations*, to be published in Papers of the Regional Science Association.

[19] S.R. Lombardo, G.A. Rabino, *Nonlinear Dynamic Models for Spatial Interaction: The Results of Some Numerical Experiments*, paper presented to the 23rd European Congress, Regional Science Association, Poitiers (1983).

[20] G. Haag, *Services 2 - A Master Equation Approach*, in C.S. Bertuglia, G. Leonardi, A.G. Wilson (eds.): *Urban Systems: Design for an Integrated Dynamic Model*, forthcoming (1987).

[21] G. Haag, A.G. Wilson, *A Dynamic Service Sector Model - A Master Equation's Approach with Prices and Land Rents*, Working Paper 447, University of Leeds (1986).

[22] G. Leonardi, *Housing 3 - Stochastic Dynamics*, in C.S. Bertuglia, G. Leonardi, A.G. Wilson (eds.): *Urban Systems: Design for an Integrated Dynamic Model*, forthcoming (1987).

[23] H. Haken, *Advanced Synergetics*, Springer Ser. Synergetics, Vol. 20 (Springer, Berlin, Heidelberg, New York, 1983).

[24] M. Clark, A.G. Wilson, *The Dynamics of Urban Spatial Structure: Progress and Problems*, Journal of Regional Science, **13**, 1-18 (1983).

[25] M. Birkin, A.G. Wilson, *Some properties of Spatio-Structural-Economic-Dynamic Models*, Working Paper 440, School of Geography, University

of Leeds (1985).

[26] P. Frankhauser, *Entkopplung der stationären Lösung des Haag- Wilson Modells durch einen neuen Ansatz für die Nutzen-Funktion*, Arbeitspapier, 2. Institut für Theoretische Physik (1987).

[27] G. Haag, P. Frankhauser, *A Stochastic Model of Intraurban Supply and Demand Structures*, in H.J.P. Timmermans et. al. (eds): *Contemporary Developments in Quantitative Geography*, (Reidel, 1988).

Chapter 8

[1] P. Nijkamp, U. Schubert, *Structural Change in Urban Systems*, Collaborative Paper CP-83-57, IIASA, Laxenburg (1983).

[2] Å.E. Andersson, *Structural Change and Technological Development*, Regional Science and Urban Economics, **11**, 267-268 and 351-361 (1981).

[3] Å.E. Andersson, J. Mantsinen, *Mobility of Resources, Accessibility of Knowledge, and Economic Growth*, Behavioural Science, **25**, 5 (1980).

[4] Å.E. Andersson, G. Haag, *Structural Change and Technological Development*, Proc. of *Research and Development, Industrial Change and Economic Policy*, Symposium at the University of Karlstadt (1987).

[5] W. Weidlich, G. Haag, *Concepts and Models of a Quantitative Sociology, The Dynamics of Interacting Populations*, Springer Series of Synergetics, **14** (Springer, Berlin, Heidelberg, New York, 1983).

[6] G. Haag, W. Weidlich, G. Mensch, *The Schumpeter Clock*, in D.Batten, J.Casti, B.Johansson (eds.): *Economic Evolution and Structural Adjustment*. Lecture Notes in Economics and Math. Systems, **293** (1987).

[7] M.S. Pena-Taveras, A.B. Cambell, *Non.Linear, Stochastic Model for Energy Investment in the U.S. Manufactoring Sector*, in R.A. Gaggioli: *Analysis of Energy Systems - Design and Operation*, The American Society of Mechanical Engineers, AES-Vol.1, 39-48 (1985).

[8] N.D. Kondratief, *Die Langen Wellen der Konjunktur*, Archiv für Socialwissenschaft und Sozialpolitik 56: 573-609 (1926).

[9] J.A. Schumpeter, *Business Cycles* (McGraw-Hill,New York, 1939).

[10] G. Bianchi, G. Bruckmann, J. Delbeke, T. Vasko (eds.), *Long Waves, Depression, and Innovation: Implications for National and Regional Economic Policy*, Proc. of the Siena/Florence Meeting (26-30 Oct. 1983), Collaborative Paper, CP-85-9, Laxenburg (1985).

[11] T. Vasko (ed.), *The Long-Wave Debate, Selected Papers* (Springer, Berlin, Heidelberg, New York, 1987).

[12] J. Delbeke, *Long-Wave Research: The State of the Art, Anno 1983*, in G. Bianchi, G. Bruckmann, J. Delbeke, T. Vasko (eds.), Collaborative Paper, CP-85-9, Laxenburg (1985).

[13] G. Bruckmann, *The Long-Wave Debate,* in G. Bianchi, G. Bruckmann, J. Delbeke, T. Vasko (eds.), Collaborative Paper, CP-85-9, Laxenburg (1985).

[14] G. Mensch, *Das Technologische Patt* (Frankfurt, 1975).

[15] G. Mensch, W. Weidlich, G. Haag, *Outline of a Formal Theory of Long-Term Economic Cycles,* in T. Vasko (ed): *The Long-Wave Debate, Selected Papers* (Springer, Berlin, Heidelberg, New York, 1987).

[16] G. Mensch, *Stalemate in Technology* (Ballinger, Cambridge, 1979).

[17] G. Bianchi, S. Casini-Benvenuti, G. Maltini, *Long-Waves and Regional Take-Offs in Italy and Great Britain: Preliminary Investigations into Multiregional Disparities of Development*, in T. Vasko (ed.): *The Long-Wave Debate, Selected Papers* (Springer, Berlin, Heidelberg, New York, 1987).

[18] M. Fischer, *Innovation, Diffusion and Regions*, in Proc. of *Research and Development, Industrial Change and Economic Policy*, Symposium at the University of Karlstadt (1987).

[19] M. Wiseman, *Economic Expansion and Establishment Growth on the Periphery*, in Proc. of *Research and Development, Industrial Change and Economic Policy*, Symposium at the University of Karlstadt (1987).

[20] J.W. Forrester, *Growth Cycles*, De Economist, **125**, 525-543 (1977).

[21] P. Nijkamp (ed.), *Technological Change, Employment and Spatial Dynamics*, Lecture Notes in Economics and Mathematical Systems, **270** (Springer, Berlin, Heidelberg, New York 1986).

[22] D. Pumain, Th. Saint-Julien, *Evolving Structure of the French Urban*

System, Urban Geography, **5**, 308-325 (1984).

[23] P. Rogerson, *The Effects of Job Search and Competition on Unemployment and Vacancies in Regional Labour Markets*, in D.A. Griffith, A.C. Lea (eds.): *Evolving Geographical Structures*, (The Hague: Martinus Nijhoff, 1983).

[24] S. Lippman, J. McCall, *The Economics of Job Search: A Survey*, Economic Inquiry, **34**, 155-189 (1976).

[25] W. A. Clark, *Structures for Research on the Dynamics of Residential Search*, in D.A. Griffith, A.C. Lea (eds.): *Evolving Geographical Structures*, (The Hague: Martinus Nijhoff, 1983).

[26] D. Mortensen, *Job Matching under Imperfect Information*, in O. Ashenfelter, J. Blum(eds.): *Evaluating the Labour Market Effects of Social Program*, (Princeton University Press, Princeton, 1973).

[27] B. Eaton, M. Watts, *Wage Dispersion, Job Vacancies and Job Search in Equilibrium*, Econometrica, **44**, 23-25 (1977).

[28] G. Haag, *A Stochastic Theory for Residential and Labour Mobility including Travel Networks*, in P. Nijkamp(ed.): *Technological Change, Employment and Spatial Dynamics*, Lecture Notes in Economics and Mathematical Systems, **270** (Springer, Berlin, Heidelberg, New York 1986).

[29] G. Haag, *Housing 2 - A Master Equation Approach*, and *Services 2 - A Master Equation Approach*, and *Labour Market 2, A Master Equation Approach for Labour Mobility,* in C.S. Bertuglia, G. Leonardi, A.G. Wilson (eds.): *Urban Systems: Design for an Integrated Dynamic Model*, forthcoming (1987).

[30] M.M.Fischer and E.Aufhauser, *Housing Choice in a Regulated Market: A Nested Multinomial Logit Analysis*, Geographical Analysis, **30**, 47-69 (1988)

[31] G.Maier and M.M.Fischer, *Random Utility Modelling and Labour Supply Mobility Analysis*, Twenty-Fourth European Congress of the Regional Science Association (1986)

[32] J.Rouwendal, Discrete Choice Models and Housing Market Analysis, University Amsterdam, Academisch Proefschrift (1988)

Chapter 9

[1] W. Weidlich, G. Haag, *Concepts and Models of a Quantitative Sociology, The Dynamics of Interacting Populations*, Springer Ser. Synergetics, Vol.14 (Springer, Berlin, Heidelberg, New York, 1983).

[2] H. Haken, *Synergetics, an Introduction*, 2nd. ed., Springer Series Synergetics, Vol.1 (Springer, Berlin, Heidelberg, New York, 1977).

[3] H. Haken, *Advanced Synergetics, Instability Hierarchies on Self-Organizing Systems and Devices*, Springer Series in Synergetics, Vol.20 (Springer, Berlin, Heidelberg, New York, 1983).

[4] R. L. Stratonovich, *Topics in the Theory of Random Noice*, Vol. 1 and 2 (Gordon and Breach, New York 1963 and 1967).

[5] A.T. Bharucha-Reid, *Elements of the Theory of Markov Processes and their Applications* (Mc-Graw-Hill, New York 1960).

[6] N. Wax (ed.), *Selected Papers on Noise and Stochastic Processes* (Dover, New York 1954).

[7] C. W. Gardiner, *Handbook of Stochastic Methods for Physics, Chemistry and the Natural Sciences*, Springer Ser. Synergetics, Vol. 13 (Springer, Berlin, Heidelberg, New York 1983).

[8] N.G. van Kampen, *Stochastic Processes in Physics and Chemistry* (North-Holland, Amsterdam 1981).

[9] G. Haag, W. Weidlich, P. Alber, *Approximation Methods for Stationary Master Equations*, Z. Physik B **26** , 207-215 (1977).

[10] G. Haag, *Transition Factor Method for Discrete Master Equations and Application to Chemical Reactions*, Z. Physik B **29**, 153-159 (1978).

[11] G. Haag, P. Hänggi, *Exact Solution of Discrete Master Equations in Terms of Continued Fractions*, Z. Physik B **34**, 411-417 (1979).

[12] P. Hänggi, G. Haag, *Continued Fraction Solutions of Discrete Master Equations not Obeying Detailed Balance*, Z. Physik B **39**, 269-279 (1980).

[13] W. Weidlich, *On the Structure of Exact Solutions of Discrete Master Equations*, Z. Physik B **30**, 345 (1978).

[14] H. Haken, *The Stationary Solution of the Master Equation for Detailed

Balance, Phys. Letters **46** A, 7 (1974).

[15] A. de Palma, *Individual Decision-Making in Dynamic Collective Systems, Part 1 and Part 2*, Journal of Mathematical Sociology, forthcoming (1987).

[16] A. de Palma, Cl. Lefe'vre, *Simplification Procedures for a Probabilistic Choice Model*, Journal of Mathematical Sociology, **8**, 43-60 (1981).

[17] T. Domencich, D. McFadden, *Urban Travel Demand: A Behavioural Analysis*, (North Holland, Amsterdam, 1975).

[18] G. Leonardi, *Transient and Asymptotic Behaviour of a Random Utility Based Stochasitic Search Process in Continuous Space and Time*, Collaborative Paper, WP-83- , Laxenburg (1983).

[19] W. Weidlich, G. Haag (eds.), *Interregional Migration, Dynamic Theory and Comparative Analysis* (Springer, Berlin, Heidelberg, New York, 1988).

Subject Index